Stars and their spectra

STARS AND THEIR SPECTRA

An Introduction to the Spectral Sequence

JAMES B. KALER

Professor of Astronomy
University of Illinois

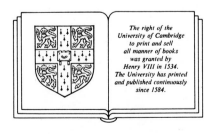

The right of the
University of Cambridge
to print and sell
all manner of books
was granted by
Henry VIII in 1534.
The University has printed
and published continuously
since 1584.

CAMBRIDGE UNIVERSITY PRESS

CAMBRIDGE
NEW YORK   NEW ROCHELLE
MELBOURNE   SYDNEY

Published by the Press Syndicate of the University of Cambridge
The Pitt Building, Trumpington Street, Cambridge CB2 1RP
32 East 57th Street, New York, NY 10022, USA
10 Stamford Road, Oakleigh, Melbourne 3166, Australia

© Cambridge University Press 1989

First published 1989

Printed in Great Britain at
the University Press, Cambridge

British Library cataloguing in publication data

Kaler, James B.
Stars and their spectra
1. Star. Spectra.
I. Title
523.8'7

Library of Congress cataloguing in publication data

Kaler, James B.
Stars and their spectra.
Includes index.
1. Stars – Spectra.   I. Title.
QB871.K33 1989   523.8'7   88-9533

ISBN 0 521 30494 6

To my Mother- and Father-in-law,
my 'other parents,'
Belle and Tibor Grossman

Contents

Illustrations

Tables and displays

Tables

Displays

Acknowledgements

I could not have completed such a wide-ranging project as this one without a great deal of encouragement, criticism, and help. First and foremost I would like to thank the editors and staff of *Sky and Telescope*, especially Ron Schorn, for their continuous support, and W. W. Morgan for very appreciated encouragement, as well as for his commentary on portions of the manuscript.

I am deeply indebted to those who read all or part of the first draft of the manuscript, who gave me suggestions to pursue, corrected scientific errors, and critiqued (and improved) my writing. I give my thanks to astronomers and colleagues Helmut Abt, William Bidelman, Anne Cowley, Art Cox, Catherine Garmany, Icko Iben, Hollis Johnson, Philip Keenan, Karen Kwitter, Julie Lutz, Dick Shaw, Harry Shipman, Jim Truran, and Ken Yoss, and to my demanding non-technical readers, friend and scholar David Bright, my mother Hazel (Susie) Kaler, and my always encouraging and supporting wife, Maxine Kaler.

I also thank those who graciously provided me with illustrations, in particular Tom Bolton, Y.-H. Chu, D. DiCicco, Lanie Dickel, Jay Gallagher, R. and R. Griffin, Bill Hartkopf, David Healy, George Jacoby, H. A. McAlister, Freeman Miller, A. G. Millikan, A. E. Morton, Ed Olson, Allan Sandage, Nolan Walborn, and Ken Yoss, and those who patiently answered my many questions. I hope you will all find your contributions within and that you will be pleased with what I have made of them.

1 *Stars*

Anyone who studies astronomy must soon encounter the ubiquitous and seemingly mystical series of letters that divides the stars into their seven groups, OBAFGKM. This alphabet of stellar astronomy, the sequence of stellar spectral types, is the foundation that supports our organized knowledge of the stars. What does this basic sequence of letters mean, where did it originate, and why is it so valuable to us? How does it serve to differentiate one kind of star from another? This book will explore these and other questions, both from a historical point of view, and from the perspective of modern observational and theoretical astrophysics.

No matter how carefully we otherwise observe stars in the sky, we cannot begin to know or understand them until we can comprehend their spectra, through which their real physical natures are revealed. We will, a pair of chapters below, set forth an overview of the subject of spectral classification, looking at its history and development, its physical basis, and at the necessary and sometimes complex embellishments that the original sequence of letters has gathered over the years. Following chapters will consider each spectral type in order, from cool M to hot O. There, we will single out prominent stellar members, explore their spectra in detail, investigate important subclasses, look at notable discoveries related to the class, and inquire into significant unsolved problems. Next we will consider the unusual stars and those that are not readily typed. At the end, we will once again tie all the classes together, and see how stars may develop from one type to another, or even pass through them all, under the inexorable forces of stellar evolution.

But before embarking on that fascinating journey, we should examine some of the fundamentals of stars, what they are like, and how they are constructed and organized. We then must delve into the sub-microscopic world of the atoms, to see how they are constructed and how they form the spectra of the stars that are the main subjects at hand. With this introduction, let us now look into the night and at the myriad stars that sparkle over our heads.

1.1 The natures of stars

The stars appear to be the primary depository of mass in the Universe. They may look like simple points of light to our eyes, but in reality, like our Sun they are vast nuclear furnaces that are capable of converting matter into energy, as a by-product

1

Figure 1.1. The classic seven-star figure of Orion, the hunter, a glorious apparition of northern hemisphere winter, at the western fringe of the Milky Way. This eminently recognizable constellation, visible at least in part from anywhere on Earth, epitomizes astronomy. Most of the bright stars are hot and blue. Rigel, at the lower right is a blue supergiant; however, Betelguese at upper left is a red supergiant that does not show up well on the blue-sensitive photographic plate used to make the picture. The larger diffuse area at lower center is the great Orion Nebula, a huge cloud of interstellar gas associated with recent star formation. From *An Atlas of the Milky Way* by F. E. Ross and M. R. Calvert, University of Chicago Press, Chicago, 1934.

illuminating our Earth, making it possible for us to exist and in turn to appreciate their beauty.

The nearest of them, the Sun, is average and typical in all sorts of ways. It is roughly one and one half million kilometers across, large enough to contain *one million* Earths. Its mass is a bit over 10^{33} grams, about a third of a million times that of our planet. At 150 million kilometers distance, its 6000 K surface sends us enough radiation to heat terrestrial water to the liquid point and make all life possible. (Degrees Kelvin,

K, are centigrade degree intervals above absolute zero, which is $-273\,°C$. Consequently, water freezes at $+273\,K$ and boils at $373\,K$). Although its light derives from nuclear fusion reactions that together act like a remarkably well controlled hydrogen bomb, we can see none of that taking place. The energy is generated in the deep interior, and what we observe has taken years to work its way through the vast outer envelope to the surface.

Although the stars in general take on a supremely unchanging mein, they are actually constantly undergoing steady alterations as their fuel supplies are used up. We are fooled because human existence is so short compared with the times over which stars live: 10 billion years for the Sun. Our own illuminating star, like most others, is in a particularly stable stage of its life, one called the 'main sequence.' But other stars have passed through this great period of stability, and are beginning to die, which causes enormous, though still slow, changes to occur. The result, coupled with nature's penchant for producing stars over a considerable range of masses, is an array of bodies with vastly different characteristics.

Let us look at some of these features. The most important physical quantity possessed by a star is its mass, which largely foretells the entire road over which it will wander during its life, as well as the age it will reach before it dies. The Sun is reasonably in the middle of a distribution that runs from roughly one tenth to 100 times the solar mass. For normal main sequence stars surface temperatures range from near $3000\,K$, for the coolest, to nearly $50\,000\,K$; and values for certain odd dying stars even reach into the hundreds of thousands. The low-mass variety radiates feebly, none visible to the eye at all, with energy outputs that are a ten thousandth of solar; at the high end, however, we see extraordinary beacons a million times brighter than the Sun that are easily visible over vast intergalactic distances. The disparity in dimensions is even greater. We see shrunken, dead stars that are comparable to Earth in size (and some bizarre objects no larger than a small city), up to swollen giants that are in the process of their death throes and that would nearly fill our entire solar system! Even the ages differ, with the five billion year old Sun again in the middle. At one extreme we know of some nearly thrice that age, and at the other, stars that were born literally (in an astronomical sense) yesterday.

This great variety is surprisingly comprehensible to us however. Over the past century we have been able to codify stars, their characteristics, and their present places in the scheme of stellar evolution. Our tool is the spectral sequence, to which we will return further along.

1.2 Names

Star names at first seem to present a picture of chaos: a number of systems are used, and any given star might be called by a variety of terms. Since we will frequently be discussing specific stars, a brief overview of nomenclature for the person new to this subject is certainly in order. Most of the brighter ones are still called by their old proper names, which issue from a variety of languages (chiefly Arabic) and are often characteristic of the appearance of the star or of its placement within its constellation.

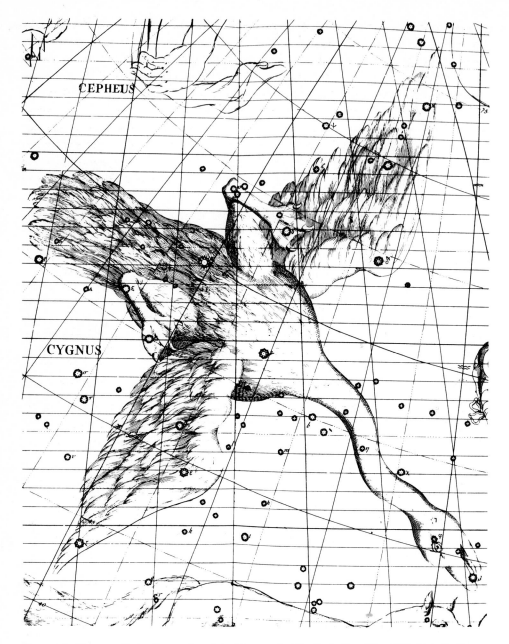

Figure 1.2. The constellation Cygnus from John Flamsteed's *Atlas Coelestis*, edition 1781. Note the Bayer Greek letters, and some Roman letters from other, now mostly defunct, naming schemes. The famous Flamsteed numbers were never added to his Atlas: Deneb is the brightest star, located just to the left of center. Courtesy University of the Illinois Library.

For example, 'Deneb' means 'tail' to the Arabs, appropriate to its end position in the constellation Cygnus the swan (Figure 1.2). These names for the twenty or so brightest stars are common in the professional astronomical literature.

A more systematic method, developed by Johann Bayer around 1600, uses Greek letters indicative of the order of brightness within a constellation, attached to the Latin genetive of the constellation name. Thus Deneb, the brightest star in Cygnus, becomes α of Cygnus, or α Cygni. The constellation name is frequently abbreviated, so that our star now becomes α Cyg. Table 1.1 gives a reference list of the names, meanings, genetives, and abbreviations of the 88 officially recognized figures, and Table 1.2 a compendium of information on the 23 brightest stars, most of which we will refer to later.

Table 1.1. *The Constellations*

Asterisks denote the 'Ancient 48', whose origins are lost in antiquity. The others are of modern origin, invented in the seventeenth and eighteenth centuries to fill in relatively blank areas of the sky and to give names to prominent southern groups not visible from the latitudes of our ancient predecessors. Some prominent groups are actually parts of larger constellations: for example the Big and Little Dippers belong to Ursa Major and Ursa Minor.

Latin name	Genitive	Abbreviation	English name
*Andromeda[a]	Andromedae	And	Andromeda
Antlia	Antliae	Ant	Air Pump
Apus	Apodis	Aps	Bird of Paradise
*Aquarius[b]	Aquarii	Aqr	Water Carrier
*Aquila	Aquilae	Aql	Eagle
*Ara	Arae	Ara	Altar
*Aries[b]	Arietis	Ari	Ram
*Auriga	Aurigae	Aur	Charioteer
*Bootes	Bootis	Boo	Herdsman
Caelum	Caeli	Cae	Graving Tool
Camelopardalis	Camelopardalis	Cam	Giraffe
*Cancer[b]	Cancri	Cnc	Crab
Canes Venatici	Canum Venaticorum	CVn	Hunting Dogs
*Canis Major	Canis Majoris	CMa	Larger Dog
*Canis Minor	Canis Minoris	CMi	Smaller Dog
*Capricornus[b]	Capricorni	Cap	Goat
*Carina[c]	Carinae	Car	Keel
*Cassiopeia[a]	Cassiopeiae	Cas	Cassiopeia
*Centaurus	Centauri	Cen	Centaur
*Cepheus[a]	Cephei	Cep	Cepheus
*Cetus	Ceti	Cet	Whale
Chamaeleon	Chamaeleontis	Cha	Chameleon
Circinus	Circini	Cir	Compasses
Columba	Columbae	Col	Dove

(continued)

Table 1.1. (*Continued*)

Latin name	Genitive	Abbreviation	English name
Coma Berenices	Comae Berenices	Com	Berenice's Hair
*Corona Australis	Coronae Australis	CrA	Southern Crown
*Corona Borealis	Coronae Borealis	CrB	Northern Crown
*Corvus	Corvi	Crv	Crow
*Crater	Crateris	Crt	Cup
Crux	Crucis	Cru	Cross
*Cygnus	Cygni	Cyg	Swan
*Delphinus	Delphini	Del	Dolphin
Dorado	Doradus	Dor	Swordfish
*Draco	Draconis	Dra	Dragon
*Equuleus	Equulei	Equ	Little Horse
*Eridanus	Eridani	Eri	River
Fornax	Fornacis	For	Furnace
*Gemini[b]	Geminorum	Gem	Twins
Grus	Gruis	Gru	Crane
*Hercules[a]	Herculis	Her	Hercules
Horologium	Horologii	Hor	Clock
*Hydra	Hydrae	Hya	Water Serpent
Hydrus	Hydri	Hyi	Water Snake
Indus	Indi	Ind	Indian
Lacerta	Lacertae	Lac	Lizard
*Leo	Leonis	Leo	Lion
Leo Minor	Leonis Minoris	LMi	Smaller Lion
*Lepus	Leporis	Lep	Hare
*Libra	Librae	Lib	Scales
*Lupus	Lupi	Lup	Wolf
Lynx	Lyncis	Lyn	Lynx
*Lyra	Lyrae	Lyr	Lyre
Mensa	Mensae	Men	Table Mountain
Microscopium	Microscopii	Mic	Microscope
Monoceros	Monocerotis	Mon	Unicorn
Musca	Muscae	Mus	Fly
Norma	Normae	Nor	Ruler
Octans	Octantis	Oct	Octant
*Ophiuchus	Ophiuchi	Oph	Serpent Carrier
*Orion[a]	Orionis	Ori	Orion
Pavo	Pavonis	Pav	Peacock
*Pegasus[a]	Pegasi	Peg	Pegasus
*Perseus[a]	Persei	Per	Perseus
Phoenix	Phoenicis	Phe	Phoenix
Pictor	Pictoris	Pic	Easel
*Pisces[b]	Piscium	Psc	Fishes

Table 1.1. (*Continued*)

Latin name	Genitive	Abbreviation	English name
*Piscis Austrinus	Piscis Austrini	PsA	Southern Fish
*Puppis[c]	Puppis	Pup	Stern
Pyxis[c]	Pyxidis	Pyx	Mariner's Compass
Reticulum	Reticuli	Ret	Net
*Sagitta	Sagittae	Sge	Arrow
*Sagittarius[b]	Sagittarii	Sgr	Archer
*Scorpius[b]	Scorpii	Sco	Scorpion
Sculptor	Sculptoris	Scl	Sculptor's Tools
Scutum	Scuti	Sct	Shield
*Serpens[d]	Serpentis	Ser	Serpent
Sextans	Sextantis	Sex	Sextant
*Taurus[b]	Tauri	Tau	Bull
Telescopium	Telescopii	Tel	Telescope
*Triangulum	Trianguli	Tri	Triangle
Triangulum Australe	Trianguli Australis	TrA	Southern Triangle
Tucana	Tucanae	Tuc	Toucan
*Ursa Major	Ursae Majoris	UMa	Larger Bear
*Ursa Minor	Ursae Minoris	UMi	Smaller Bear
*Vela[c]	Velorum	Vel	Sails
*Virgo[b]	Virginis	Vir	Virgin
Volans	Volantis	Vol	Flying Fish
Vulpecula	Vulpeculae	Vul	Fox

[a] In Greek mythology Andromeda was the daughter of Cepheus and Cassiopeia and wife of Perseus; Cassiopeia was the wife of Cepheus; Cepheus was a king of Ethiopia; Hercules was a son of Zeus and celebrated for his great strength; Orion was a hunter who chased the Pleiades and was slain by Diana; Pegasus was a winged horse; and Perseus was a son of Zeus and was the hero who rescued and married Andromeda.
[b] Constellation of the Zodiac, the band that contains the ecliptic, the apparent annual path of the Sun.
[c] Carina, Puppis, and Vela are subdivisions of the original very large constellation Argo Navis, the legendary ship Argo. The modern constellation Pyxis is sometimes considered to be a part of it.
[d] Contains two parts separated by Ophiuchus: Serpens Cauda, the head, is the eastern half, and Serpens Caput, the tail, is the western. It is still considered to be one constellation.

Table 1.2. *The First Magnitude Stars*

The bright stars below are ordered according to apparent visual magnitude, V. Adhara and Castor, the brightest of the second magnitude stars, and the Sun are also included. The list gives the B–V color index, the distance in parsecs (D), the absolute visual magnitude (M_v), the spectral type (Chapter 3), and some special remarks. Distances under 25 parsecs are derived from parallax (Chapter 5), which are then used to derive M_v. Absolute magnitudes and distances for stars farther than 25 parsecs are inferred from the spectral class (again see Chapter 5). Separate binary components are listed for α Cen and α Cru. Several of the stars have faint companions that are not indicated here.

Bayer name	name	V	B–V	D (pc)	M_v	Spectrum	Remarks
—	Sun	−26.74	0.65	—	4.83	G2 V	
α Canis Majoris	Sirius	−1.46	0.00	2.65	1.42	A1 V	Binary with faint companion.
α Carinae	Canopus	−0.72	0.15	70	−5	F0 II	
α Centauri	Rigel Ken-	−0.01	0.71	1.33	4.37	G2 V	Combined naked eye
	taurus	1.33	0.88	1.33	5.71	K1 V	mag. $V = −0.29$. Tenth mag. companion, Proxima.
α Bootis	Arcturus	−0.04	1.23	10.3	−0.10	K1 III	
α Lyrae	Vega	0.03	0.00	7.5	0.65	A0 V	Standard star for astronomical photometry.
α Aurigae	Capella	0.08	0.80	12.5	−0.40	G5 III	Visually unresolved
						G0 III	binary with roughly equal components.
β Orionis	Rigel	0.12	−0.03	265	−7	B8 Ia	
α Canis Minoris	Procyon	0.38	0.42	3.4	2.71	F5 IV	Binary with faint companion.
α Eridani	Achernar	0.46	−0.16	27	−1.7	B3 V	
α Orionis	Betelgeuse	0.50	1.85	320	−7	M2 Ia	
β Centauri		0.61	−0.23	95	−4.3	B1 III	
α Aquilae	Altair	0.77	0.22	5.0	2.30	A7 V	
α Tauri	Aldebaran	0.85	1.54	19	−0.49	K5 III	
α Scorpii	Antares	0.96	1.83	190	−5.4	M1.5 Ib	Binary with faint hot companion.
α Virginis	Spica	0.98	−0.23	67	−3.2	B2 V	Binary with roughly equal components.
α Crucis		1.58	−0.26	120	−3.8	B0.5 IV	Binary with com-
		2.09			−3.3	B1 V	bined naked-eye magnitude of $V = 1.05$.
β Geminorum	Pollux	1.14	1.00	10.6	1.00	K0 III	
α Piscis Austrinus	Fomalhaut	1.16	0.09	6.7	2.02	A3 V	
α Cygni	Deneb	1.25	0.09	500	−7.2	A2 Ia	
β Crucis		1.25	−0.23	150	−4.6	B0.5 III	
α Leonis	Regulus	1.35	−0.11	22	−0.38	B7 V	
ε Canis Majoris	Adhara	1.50	−0.21	190	−4.9	B2 II	
α Geminorum	Castor	1.59	0.03	15	0.72	A1 V	Multiple star (see Section 1.6).

The brightness ordering is not strictly observed: β Orionis is brighter than α, and the stars of Ursa Major are ordered by their positions in the Big Dipper. More importantly, there are only a limited number of letters, although the system is extended somewhat by the use of numerical superscripts for neighboring stars, e.g. π^1, π^2, π^3, π^4, and π^5 Orionis. A more comprehensive method derives from John Flamsteed about a hundred years later, who ordered stars from west to east within the constellation. Thus Deneb = α Cyg also becomes 50 Cyg. The Bayer letters almost always take precedence, however.

When the Flamsteed numbers are not available (and even when they are for faint stars) we rely on numbers from catalogues that are divorced from the constellations: the *Bright Star Catalogue*, which includes all the naked-eye stars and more (Deneb = HR 7924, where HR stands for 'Harvard Revised'); the Henry Draper spectroscopic catalogue, which we will encounter extensively later (Deneb = HD 197345); the Bonner Durchmusterung (Bonn Survey, Deneb = BD or DM +44°3541); or the relatively new Smithsonian Astrophysical Observatory compendium (Deneb = SAO 049941). A variety of specialty names will also arise as we proceed. So, as we see, things are not so disordered after all.

1.3 Locations

We also indicate and organize the stars by prescribing a pair of coordinates. The most common version is a simple analogue of terrestrial latitude and longitude. We define celestial poles and a celestial equator that lie above the Earth's rotation poles and equator. We next establish a fundamental meridian on the sky, akin to the prime meridian on Earth, that runs between the poles and through the Vernal Equinox, the basic celestial reference point where the Sun crosses the equator on its way north. We then pass another meridian, called an *hour circle*, through the star whose coordinates are to be specified. The angle between the Vernal Equinox and the point where the hour circle crosses the celestial equator is called the *right ascension* (α) of the star, and is usually measured in time units (hours, minutes, and seconds) where 15° equal one hour (h), one degree equals 4 minutes (m), etc. The angle from the equator to the star, measured north or south (+ or −) in degrees along the hour circle, is called *declination* (δ). As an example, the coordinates of Deneb are

$$\alpha = 20 \text{ h } 41 \text{ m } 25.8 \text{ sec} \qquad \delta = +45° \, 16' \, 49''.$$

Because of precession, a wobble of the Earth's axis, the coordinates change with time, so that we must also specify a date (called the *epoch*) for which they are exactly correct. For the position of Deneb above the epoch is the beginning of the year 2000. Since we know how α and δ vary we can easily calculate what they should be for any moment in question. In the absence of names for stars or other celestial objects, we may then simply specify the precise coordinates. The '+44 °' in the BD number of Deneb above refers to a strip of declination in which the star was found in 1875, the epoch used for the catalogue, which pointedly demonstrates the importance of this motion.

1.4 Distances

Our view of the nighttime sky is very deceptive. We see the stars crowded together and in clear mountain air we think that we might even pluck one down to Earth. The distance of a star is probably its most fundamental observational quantity, since without it we can derive little else. The values are enormous compared with anything we are used to, even that for the Sun, which seems so far away, and their acquisition has been a continuing quest in astronomy for the past century and a half. As we proceed through the tale of this volume we will introduce and discuss many methods; here, we simply look at results, so that the novice to astronomy can get some feel for the context of the subject.

In any science, we would prefer to use measurement units that are appropriate to the sizes of the things measured. For example in the solar system, we do not use kilometers, but the *astronomical unit* (AU), the average distance between the Earth and Sun (technically, the semi-major axis of the Earth's elliptical orbit). In stellar astronomy, especially in popular or semi-popular accounts, we might use the *light-year*, the distance a ray of light or photon travels in a year, which immediately gives us some notion of scale. The light-year (l-y) is 63 300 AU or 9.53×10^{12} kilometers long. The *nearest* star, α Centauri (a gravitational companion labeled *Proxima* is actually closest to us) is 4.3 l-y, or just over 270 000 AU away. This separation is typical. Since the naked-eye sky is dominated by intrinsically luminous stars, many that we see are much farther: the middle Dipper stars are roughly 100 l-y distant, and our previous example, Deneb, some 1600! The light we see from α Cyg left the stellar surface in the year 400.

We can put these distances into better perspective if we compare them with the Sun and solar system. The Sun has a diameter of only 0.01 AU, so that it is separated from its nearest neighbor by *27 million times its own size*: again typical. This distance is even 6800 times the dimension of our entire planetary system. Thus we see that space is so very empty: within our local neighborhood only one part in 10^{22} is actually occupied by stars; collisions simply do not occur.

The professional astronomer uses a somewhat different unit, the *parsec* (pc), the distance at which the astronomical unit would appear to be one second of arc across. It is thus equal to 206 265 AU (the number of seconds in a radian), or 3.26 l-y. Our neighbor α Cen is therefore 1.3 pc away and Deneb 500. These units are subject to the usual multiplying prefixes: kiloparsec (kpc), megaparsec (Mpc), and gigaparsec (Gpc) for 10^3, 10^6, and 10^9 pc. We see from the latter just how far some astronomers (but in this book, not us) need go. Parsecs will be used throughout this account; since distances are generally not all that well known, just triple them to recover light-years.

1.5 Compositions

What the stars are made of, and how we learn of it, is a major theme to be developed later. But again, it is important to establish a simple framework in which we will work, so the skeletal results are presented first. To a limited approximation, the Universe is all hydrogen and helium, the simple stuff that developed out of the creation event that took place some 13 billion years ago. The cores of stars, where nuclear fusion takes place and heavier elements are built, are notable exceptions. The outer parts of the Sun and other common stars, however, follow the general rule of 90% H, 10% He (by number, not weight), and, remarkably only 0.1 % everything else! By contrast, we see that our Earth, with its domination by heavy elements, is quite bizarre: the planets are, in a sense, element-concentrating devices.

Within the 'everything else' category, we find as a (frequently broken) rule that the heavier the element the less there is of it. Usually, oxygen dominates the group, followed by carbon and nitrogen, then neon, with iron quite noticeably abundant for its weight. Within this array of elements beyond helium, we find considerable variation from one star to the next, a matter upon which we will focus in the chapters to come.

1.6 Organization: double stars

In spite of their seemingly random distribution on the darkened celestial sphere, the stars are quite highly organized. First, we find that they prefer to form doubles (called *binaries*) and multiples: perhaps necessary to the formation process itself. Most of the naked eye stars have some sort of companion, and fully 20% of those that appear to be single will separate into a double with the use of a modest telescope. Sometimes the members of the pair are quite similar, other times hugely disparate. The brightest star in the Southern Cross, α Crucis, consists of a closely-spaced but still resolvable binary, whose constituents are almost perfect twins. Near the other extreme might be Sirius, the brightest star in the sky, with a companion that is almost 200 000 times feebler than the great naked-eye component. Extending ourselves further, we might even include our Sun and Jupiter in the pantheon of doubles. The planet is not a star only because its mass is too low, and it does not generate its own energy. The majority of doubles are too close together to be seen optically, that is directly with the telescope, and we must rely on more sophisticated techniques, including spectroscopy (the science of the analysis of light by its color, or wavelength: see Chapter 2), to detect them.

These stars are of immense value, as they are the sources of our knowledge of stellar masses, which are derived from the characteristics of the orbits as the bodies swing about one another. If we watch a visually observed binary (one that can be seen as a double by eye alone) for long enough – perhaps many years or decades – we can plot the orbit of one star about the other, and determine its orbital period, P. If we know the distance, we can calculate the physical size of the path, a (actually the semi-major axis of the elliptical orbit). We can then apply Johannes Kepler's third law of planetary motion, as generalized by Isaac Newton, which relates P, a, and the sum of the masses of the two stars: $P^2 = 4\pi^2 a^3/G(M_1 + M_2)$, where G is the universal constant of gravitation, and M represents a stellar mass.

In any orbital pair, however, one body does not exactly go around the other; each revolves about a common *center of mass* that lies on a line somewhere between the two. Its distance from either body is inversely related to the ratio of masses. In the case of the Earth and Moon, for example, the center is 80 times closer to the former than it is to the latter, because our planet has 80 times the mass of its companion. Consequently, the Earth orbits a point that is $\frac{1}{80}$ the distance to the Moon, or 5000 kilometers from its center, still 2000 kilometers below its surface. If by careful observation we can locate the center of mass, we can deduce the mass ratio of the members of the binary pair. This number coupled with the sum of the masses solved from Kepler's third law yields the individual values. Similar data can be inferred from spectroscopically detected pairs and from those that eclipse each other's light as they swing through their orbits. We will encounter the binaries many times over the course of these pages.

But the doubles are far from the end of the story. Triples abound: our neighbor α Centauri appears through a small telescope as a closely spaced double, with one member about three times as bright as the other; roughly a degree away is Proxima, only about 1/10 000 the luminosity of the brightest component. Proxima circles the close, bright pair in an immensely long orbit, with a period estimated to be hundreds of thousands of years, and is now on the terrestrial side, making it the closest star. Quadruples? Look at Mizar, the next-to-last star in the handle of the Big Dipper, which breaks into a visual double with a 14 second of arc separation ('easy' for the amateur). Each of these is an even closer pair, as found with the spectrograph: a double-double. An equally famous example is ε Lyrae, in which all four can be seen at fairly high telescopic power. Mizar is actually part of a quintuple, as it has a naked-eye companion called Alcor (Figure 1.3: find it on any decent star map). Castor, α Geminorum, is an example of a sextuple: two doubles together that make the bright naked-eye star in Gemini with another one farther away.

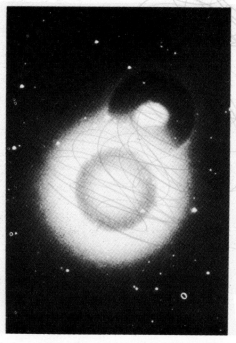

Figure 1.3. The famous naked-eye double Mizar and Alcor, ζ and 80 Ursae Majoris. Mizar is split again in a small telescope, and then each of the two is divided yet again to make a quintuple system. Lowell Observatory photograph.

1.7 Clusters

At some vague point (ten?) the multiplicity is such that the group is more properly called a cluster. There are two kinds of these. The more common, by far, are the *galactic clusters*, which hug the plane of the Milky Way, the broad band of light that girdles the sky. These range in membership from only a few stars to several hundred, and tend to be on the young side, since most of them are not very strongly bound together by mutual stellar gravity. Stars like to form in such groups. Even the Sun was probably once a part of one that long ago disintegrated. Several famous examples are visible to the naked eye, notably the Pleiades, or Seven Sisters, in Taurus, and nearby, the Hyades, centered on the bright reddish star Aldebaran, α Tauri (which is not a cluster member). In the south, look for the Jewel Box (Figure 1.4), bright enough to merit its own Greek letter as $\varkappa$ Crucis. Less well known is the barely visible Praesepae (Beehive Cluster) in Cancer. A tour through the Milky Way with just binoculars will reveal numbers of them. The Sun is actually inside a rather poor galactic cluster that consists of the middle five stars of the Big Dipper, Sirius, α Coronae Borealis and several others. It is just passing through, however, and will eventually exit in about a

Figure 1.4. The Jewel Box cluster, $\varkappa$ Crucis, an archetypal galactic cluster and one of the showpieces of the south. Thousands of these clusters, within which all the stars were born at about the same time, can be found throughout the Milky Way. Several of what appear to be individual stars are doubles, triples, and multiples. From the ESO/SRC Southern Sky Survey.

Figure 1.5. The greatest globular cluster Messier 3 in Canes Venatici, one of the best of a set of about 200 such ancient objects that inhabit the halo of our galaxy. Several hundred thousand stars are packed into a volume only a few tens of light years across. National Optical Astronomy Observatories (Kitt Peak) photograph.

million years, allowing future earthly residents to see the group move off into the distance.

Much more spectacular are the rarer *globular clusters* (Figure 1.5). These are immense systems that contain from tens of thousands to perhaps a million stars. They are not just the upper range of the galactic clusters, but are generically different, much older, with different kinds of stars than those we see around us, stars that are highly deficient in metal atoms. The best known examples are the extraordinary ω Centauri in the south, easily visible to the naked eye, and M13 (or Messier 13, named after a seventeenth century comet hunter, Charles Messier) in the north, a binocular object in Hercules.

1.8 The Galaxy

All of these – singles, doubles, multiples, and clusters – are assembled into a system called simply our Galaxy, or the Milky Way galaxy after its chief manifestation. The overall simplified structure is shown in Figure 1.6. Most of the two hundred billion stars it contains are concentrated into a thin disk some 30 000 pc (100 000 l-y) across.

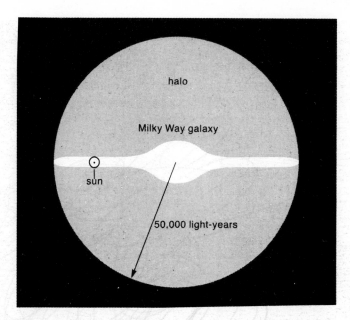

Figure 1.6. Diagram of our Galaxy, showing the thin disk that contains most of its two hundred billion stars, the Sun, interstellar gas and dust, and the galactic clusters. It is there, in what we call Population I, that new stars are being born. The halo encloses an older, earlier generation of stars called Populaton II that includes the globular clusters, and that was born before the disk developed. Diagram by the author, courtesy of *American Scientist*.

The Sun is located in the disk, about 8 kpc from the center and since we are inside it, the combined light of all its stars then appears as a band – the Milky Way – about our heads (Figure 1.7). Within this broad pathway we see large quantities of free gas and dust, out of which new stars are continuously being born, so that here we find also the relatively young galactic clusters. Our disk is slowly rotating and has broken into a spiral pattern beautifully seen in other such systems (Figure 1.8). Its stars, gas, clusters, and dust are collectively referred to as *Population I*.

Enveloping the disk is a gigantic halo of ancient stars called *Population II* that includes the globulars. The halo was the first part of the Galaxy to form and is low in heavy elements relative to the younger disk and Sun. This relation between age and composition is caused by the aging processes of the stars, which cause heavy elements to be created and then spewed back to the interstellar gas, out of which later, enriched, generations are born.

There are countless other galaxies, each, like M101 in Figure 1.8, usually distinct and well separated from the others. The nearest one to ours of comparable size is M31 (Figure 1.9), easily visible to the naked eye in Andromeda, distant by about 20 times the Milky Way system's diameter. Galaxies come in many sizes and shapes, from these spectacular spirals to mere, small featureless blobs. It is fair to say that with the exception of a few escapees, all stars are members of galaxies.

a

Figure 1.7. (*a*) The Milky Way, the disk of our Galaxy, in a spectacular appearance that runs from Cygnus, the Northern Cross, at the upper left edge to the Southern Cross at lower right, and that features the galactic center in Sagittarius. The dark line that seems to divide it in two is caused by a thin disk of nearby, obscuring dust that lies in the space between the stars. (*b*) shows a map that locates prominent stars and planets (as well as Halley's Comet!). The 'summer triangle' consists of Deneb at the left, Vega on top, and Altair at the right. Crux, the Southern Cross, is indicated by the small cross at the right. The 'SMC' is the Small Magellanic Cloud, a nearby companion galaxy. The streak across the center was caused by a passing meteor. The glow at the lower left is the *zodiacal light*, caused by dust in the solar system reflecting sunlight. Photograph and map courtesy of Dennis di Cicco and *Sky and Telescope*.

b

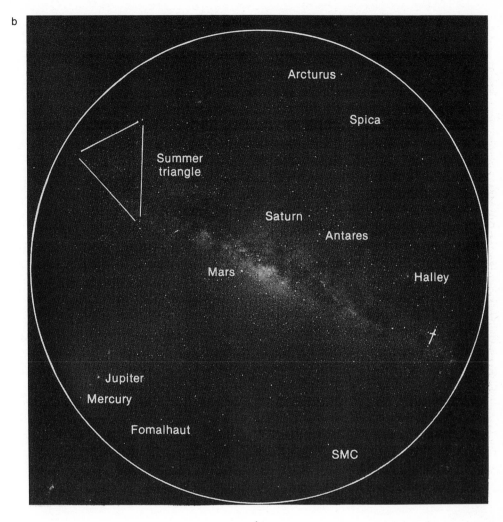

b

Figure 1.8. The spectacular galaxy M101, whose disk, presented face-on, shows a superb wide-open spiral pattern. The Milky Way galaxy's arms are more tightly wound. National Optical Astronomy Observatories photograph.

Figure 1.9. The Great Nebula in Andromeda, the spiral galaxy Messier 31. A close match to our own, this nearby galaxy, only 700 000 pc away, is presented at about a 70° angle from face on, so that one can appreciate the thinness of the disk as well as its spiral structure. The small blob at the lower edge is an elliptical-type companion. National Optical Astronomy Observatories (Kitt Peak) photograph.

1.9 Motions

We speak of the 'fixed stars,' but they are fixed only in comparison to the rapidly moving planets. Since all the stars are moving in different orbits around the center of our Galaxy, they must all drift with respect to one another. Thus from our perspective, all the stars in the sky appear to be in slow relative motion. If we could speed up time by a factor of say, a billion, it would be like wandering through the throng at a party. Individuals would move toward us, get seemingly larger (in the case of the stars, brighter), then disappear into the distance. Of course the stars are so far away that even at typical relative speeds of 30 or so kilometers per second relative to the Sun the motion is not sensible over a lifetime. But if we could return in 100 000 years, we would witness a real distortion of our constellation figures. With telescopic power, however, we can easily follow the paths taken by our neighbors.

Of course, the stars are moving around us in three dimensions. What we have described above is that component of motion that carries a star across the line of sight, a quantity measured in seconds of arc (or fractions thereof) per year or century, called the *proper motion*. This information coupled with distance yields the actual speed perpendicular to our view, the *transverse velocity*. The other component, that along the line of sight, is called the *radial velocity*, and is directly derivable from spectroscopy: we will explore its measurement further in Chapter 2. When combined, the two components give us the three-dimensional *space velocity*, which in turn provides us with information on the dynamics of the Galaxy, and tells us that our Sun is moving through the surrounding swarm roughly in the direction of the constellation Hercules.

1.10 Starlight: the electromagnetic spectrum

We now turn from looking at geometrical attributes of stars to the examination of their radiant properties. First, however, we must introduce the subject of radiation itself and the *electromagnetic spectrum*. The most familiar example of electromagnetic radiation (EMR) is *light*, energy in a form that can be readily detected by our eyes. Light can be thought of as alternating electric and magnetic fields that, in a vacuum, flow at the 'speed of light,' 300 000 kilometers per second, c. The behavior of EMR depends upon the experiment or observation: sometimes it appears as moving waves, analogous to those in a pool of water; other times its effect can only be understood by assuming that it consists of a flow of particles. In order to describe this dual nature we use the *photon*, a particle that in a loose sense acts like a chunk of the wave.

We categorize the kind of radiation we observe by either its *wavelength*, λ, the separation between wave crests within the photon, or by the *frequency*, ν (sometimes f), the number of wave crests that would pass a particular point per second. The latter is measured in cycles per second, or hertz (Hz); a thousand cycles per second is a kilohertz (kHz), a million is a megahertz (MHz) etc. The number of crests that you would count per second multiplied by their separation obviously equals the distance that the wave (or photon) travels in that interval or (wavelength) × (frequency) = (velocity), $\lambda\nu = c$.

EMR transfers energy from one place to another in the Universe. The energy (E) carried by an individual photon is directly proportional to its frequency, and inversely proportional to wavelength, or $E = \mathrm{h}\nu = hc/\lambda$, where h is a constant of nature called *Planck's* constant. In order for a beam of long-wavelength photons to carry as much energy as one of short wavelength, we must have more of them. The number in the beam times the energy of each one is called the *intensity* or *flux* depending upon what units we choose to define the flow rate. The energy of a single photon is very small. Planck's constant has units of 6.6×10^{-27} erg seconds. When multiplied by the frequency of, say, yellow light (5.5×10^{-14} Hertz), we find an energy of just 3.6×10^{-12} erg. By comparison a radiant power of one watt is 10^7 ergs per second, so that a standard hundred watt light bulb radiates some 10^{20} photons per second.

EMR exists over an immense range of wavelengths (see Figure 1.10). The array is called the *electromagnetic spectrum*, vaguely defined sections of which are given descriptive names. The centerpiece, as far as humans (and most of the subject matter of this book) are concerned, is *optical* or *visual* radiation, that which we can see with our eye, which has wavelengths between about 4×10^{-5} and 8×10^{-5} centimeter. In order to avoid exponents, we use the more common *Ångstrom unit*, where $1\,\text{Å} = 10^{-8}$ cm; we then view the world over a mere one octave between 4000 and 8000 Å.

Within this optical band, radiation of differing wavelengths is seen as color: red about 7000 Å and longer, orange near 6000, yellow, green, blue around 5500 Å, 5000 Å, and 4500 Å respectively, and finally violet shorter than ('below' in the jargon, with 'above' implying longer wavelengths) 4500 Å. Physiologically, color can also be produced by combinations of the pure spectral colors; white is a mixture of all of them; sunlight contains all wavelengths as well, but is dominated by yellow, which gives it its characteristic hue.

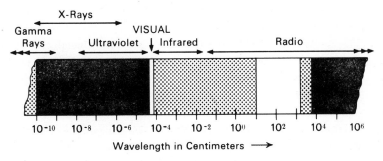

Figure 1.10. The *electromagnetic spectrum* from γ rays to radio waves. Note the very small segment that belongs to the visual, which runs from violet light at the short wavelength end through blue, green, yellow, and orange to red at the long-wave limit. The degree of shading crudely indicates the transparency of the Earth's atmosphere, through which we must look to see the stars. Only a very small part of the ultraviolet adjacent to the optical can reach ground-based telescopes. The infrared and radio are chopped up with alternating regions of high and low opacity. Adapted from *Principles of Astronomy, A Short Version*, 2nd edn., by S. P. Wyatt and J. B. Kaler, Allyn and Bacon, Boston, 1981.

If we extend our view from visual to longer wavelengths, we first encounter *infrared* (*IR*) beyond red, extending to roughly a millimeter, and then *radio*, which has no upper limit. Below the violet, also invisible to the eye, we find the *ultraviolet* (*UV*). That portion close to the optical spectrum is called 'near UV' the rest 'far UV'; these terms also apply analogously to the IR. The UV is transformed to *X-ray* under roughly 100 Å or so, and below 1 Å we refer to the photons as *gamma* (γ) rays. (More jargon: the shorter γ and X-rays within their sets are called 'hard', the others 'soft').

What we have here of course is an array of energies. Keeping in mind that E is proportional to ν or $1/\lambda$, optical photons carry 10^9 times more energy than the radio photons associated with, say, standard AM broadcasting. These longer-wave photons are (unless they come in vast numbers) harmless and pass through you without your being aware of them. No one fears the radio towers that fill our space with low-energy EMR. But photons shorter than the optical are another matter. These can produce serious damage. Because of the ozone layer, the Earth's atmosphere is opaque to radiation below 3000 Å. But the near UV between that cutoff and the 4000 Å violet rays is enough to produce sunburn. The heat you feel from the Sun is largely in the infrared, so you can still get burned without being very warm. Danger from X-rays is well publicized, and γ rays are one of the principal hazards associated with nuclear reactions.

The electromagnetic spectrum not only carries energy, it carries to us facts on the sources that emit it. Different kinds of information about temperatures, densities, compositions, and motions arrive at different wavelengths. For example much of what we know about stars is found from the optical spectrum, whereas our data on galactic rotation is derived largely from radio observations, and the knowledge of high-energy processes in galaxies comes from celestial X-rays. Consequently, astronomers need to be able to observe the entire spectrum. Our atmosphere, however, prevents us from doing so: Figure 1.10 indicates its transparency. The optical spectrum, where the Sun produces its energy, and which heats the land, is mostly passed through, as is a small portion of the near UV. But, fortunately for life on Earth, the majority of the ultraviolet and all of the damaging X-rays cannot reach the ground; only in the γ-ray region, where there is little natural radiation, can any energy penetrate. The atmosphere is partially transparent to infrared and radio. Within the IR our view of space is highly variable and dependent upon the exact wavelength. Fortunately there is a broad radio band that gives us good command of low energy processes. In order to view the heavens in the various forbidden zones, we must fly telescopes and other instruments on rockets and satellites to carry them above the air that blankets us. Most of this book is devoted to optical studies, but in numerous instances we shall incorporate data brought to us by space astronomy from the other wavelength domains.

1.11 Brightnesses: apparent magnitudes

With these fundamental definitions and descriptions in place, let us now look at the various properties of stars that describe their radiation: magnitude, color, and temperature. The fundamental observational datum that describes a star's apparent brightness as viewed from Earth is a quantity called the *apparent magnitude*. Ideally, we should probably use physical units, the amount of energy passing through a square centimeter per second (a quantity called the *flux*), which in fact we occasionally do. But astronomy is an ancient science, with a vast pedigree: we needed a measurement unit long in advance of even our concept of energy, and once a system is in place, and elaborated upon, there is little sense in changing it. Better simply to calibrate it against newer units.

Our system developed from old Greek times when the stars were divided into six brightness categories called *magnitudes* (m). Those of 'first magnitude' were brightest, while those of sixth were at the limit of human vision. In the nineteenth century, astronomers found that this system was actually a logarithmic brightness scale, and that the five magnitude divisions corresponded roughly to a factor of 100 in actual apparent luminosity. The simple procedure was then to codify the system exactly as such, so that one magnitude would be the fifth root of 100, or 2.512, and to decimalize it. According to convention, stars with magnitudes between 0.50 and 1.49 are collectively called first magnitude, those between 1.50 to 2.49 belong to the second, and so on.

The end of the tale is that sixth magnitude (specifically $m = 6.00$) was defined by the average of a set of faint stars near the limit of human vision, with the brighter ones measured against them. With this scaling, the most luminous stars become brighter than first so we had to adopt negative numbers. Alpha Centauri (the combined light of the pair) then has magnitude -0.29 (in the 'zeroth magnitude' category, and Sirius -1.46 (the minus first). The planet Venus outshines them all at -5 and the full Moon and Sun glow at -13 and -27. With telescopes, we observe fainter stars and magnitude values above sixth. With binoculars you can see eighth magnitude, a small backyard telescope shows perhaps the 11th, the large 5-meter Palomar reflector reveals the 26th, and the Space Telescope will allow us to probe to near 30th. Do not be deceived by the small numbers: remember each step is a factor of 2.5 in brightness. The great reflectors are detecting stars 20 magnitudes, or 10^8 times fainter than the human eye can see alone.

Measurement of magnitude can be done either photographically or electronically. If we expose a photographic plate to the sky, the sizes of the stellar images will be crudely proportional to brightness. However the conversion is complex and precision difficult. A far superior method is to use a *photoelectric photometer*. This device, placed at the focus of a telescope, contains a light-sensitive material that, with the aid of appropriate electronics, can convert radiative energy into an electric current. The current is easily measurable, and is directly proportional to brightness, allowing for accurate magnitude determinations. An even more advanced system is the *charge-coupled device* (CCD), which can convert the light from a whole field of stars into a flow of electrons that with the aid of a computer can be converted into a photograph-like picture, with exact magnitudes.

1.12 Variable stars

Few casual watchers of the sky are aware that stars are not necessarily constant in their brightness or magnitude. Most are, otherwise we would see the constellation patterns change from night to night. But a surprising number do show such changes.

These *variable stars* come in a great variety of types, which will be discussed and expanded upon in some detail as the subject repeatedly arises in the chapters that follow.

Most variability is intrinsic to the star itself and is caused by some sort of instability that produces pulsations and changes in the stellar diameters. We see some stars that oscillate over several magnitudes with periods of years, others with easily recognized daily variations, and still others that subtly chatter away over intervals of minutes. The collection is quite wonderful and fascinating and tells us a great deal about stellar structure and the aging process. To these groups we also must add the double stars whose orbital planes are positioned in the line of sight so that one star passes in front of, or eclipses, the other.

The brighter variables, such as the famed stars Algol (β Persei), Mira (o Ceti), δ Cephei, and β Lyrae carry ordinary proper names, Bayer letters, or Flamsteed numbers. (We will return to these individuals at an appropriate point). Otherwise, variables are designated in a distinctive fashion. The first to be discovered in a constellation (excepting those with standard names) is called R, with the Latin genitive of the constellation, as in R Cygni, the second S and so on. Presumably, not too many such stars were originally expected. After Z, follow RR (such as the well known RR Lyrae), RS to RZ, SS to SZ, and so on. After ZZ, we start with AA . . . AZ, etc., down to QZ, which (with J left out) gives us 334 combinations. Many constellations contain far more, and we finally give up the letters, and begin with V 335. The system is a bit arcane, but effective and in universal use.

1.13 Absolute magnitudes

The apparent magnitude – the brightness of a star as it appears to us on Earth – depends upon two variables: the amount of energy produced by it per second, or the *intrinsic luminosity*, and the distance. Some stars, like α Cen, are bright only because they are close to us. We need a way to register the true brightness, and again we employ magnitudes. The scheme is to use the inverse square law of radiation (which says that the apparent luminosity of a light source changes according to the square of the inverse of the distance) to array the stars mentally at a common position arbitrarily chosen to be 10 pc, so that they can be directly compared with one another. The resulting values are called *absolute magnitudes*, given by M. With the use of these, we would see some interesting changes in our view of the sky. The absolute magnitude of our brilliant Sun is only 4.8, about as bright as the faintest star of the Little Dipper, which is quite invisible from an illuminated town. Deneb on the other hand shines at a remarkable $M = -7$, so bright that at 10 pc it would cast shadows, much like Venus. We will look at the full array of stars, from brightest to faintest, which covers an astonishing 28 magnitudes, later.

Although mathematical equations are kept to a minimum in this volume, it is useful to see the conversion relation between the various quantities: $M = m + 5 - 5 \log D$ where D is in parsecs. If we have any two we can derive the third. And note here a matter of particular importance to be discussed below, that if m is measured, and M somehow inferred (which we can do), we can find that elusive value of distance.

1.14 Color

Even a quick look at the brighter stars will show differences in their colors. Serious examination, especially with a small telescope, reveals broad variations among their hues. We see red, orange, yellowish, through white to blue, in accord with the progression of colors through the spectrum, except that white substitutes for green. Occasionally, we will even find one a deep crimson. These colors are closely related to the spectral types that we will begin to examine in Chapter 3, and they are intimately connected to temperature, which we consider both below in the next section, and again later as the physical basis of spectral classification.

However, color can be treated and discussed without any reference to these parameters. The magnitude of a star has been discussed above purely from the point of view of a visual observer, one who uses, and estimates with, only his or her own eyes. The perceived brightness of a star must depend upon the energy output of the star, or rather the total energy from that star falling on the Earth, and *where in the spectrum the radiation is concentrated*. To use an extreme example, if a star radiates all its light in the infrared, to which the eye is totally insensitive, the star will be quite invisible no matter how much radiation it produces.

Real starlight, with the exception of that from the very coolest, or the most dust-obscured stars, is always a mixture of all colors, the dominant one giving the hue perceived. And in spite of the physiological color of the star, the eye is principally a yellow-sensitive detector. Therefore among blue, yellow or orange stars of equal radiant energy, we will see the yellow one as brightest, and the visual magnitude scale will be biased toward the amount of that color in the mix.

The first quantitative measurements of magnitude were derived from photography. The photographic plate, however, is sensitive primarily to blue light and will therefore order brightness more according to the amount of *that* color. The original practice was to photograph the sky with a normal blue-sensitive plate and then with one that is used in conjunction with a filter to mimic the color response of the human eye. A magnitude scale was established for each system, photographic (m_{ptg}) and photovisual (m_{vis}) respectively, with the numerical values set to be the same for white stars. The observers then defined a *color index*, CI, which is the algebraic difference between the two, $m_{ptg} - m_{vis}$. Since blue stars will appear brighter photographically, m_{ptg} will be lower than m_{vis}, and CI will be negative, and vice versa for red or yellow stars. Red Betelgeuse, visually striking and first magnitude to the eye, shrinks to a weak second to the unfiltered blue-sensitive camera (refer back to Figure 1.1).

As pointed out above, the photographic plate is a poor instrument for quantitative measurement, as it is quite non-linear (that is, the response to a specific change in signal depends on the strength of the signal); electronic devices are linear and far superior. In order to place these newer measurements onto the scales of the older system, photoelectric measurements employ filters to mimic both the effect of the eye and the photographic plate. There is then no reason why we must be limited to two colors. The most common photoelectric magnitude system uses *three* colors, with an ultraviolet sequence (U) added to those in the blue (B, similar to m_{ptg}) anf visual (V, similar to m_{vis}). Note that the letter '*m*' is no longer used in this system. All three magnitudes are again equal for white stars, and we now have two color indices, U − B and B − V. The latter, the most commonly used, ranges from −0.4 magnitudes for blue stars to about +2.0 for the deeper reds.

We need not stop there: red and infrared magnitudes have been added at longer wavelengths, yielding the UBVRI system, with even longer-wave measurements added at positions called J, K, etc. These are especially useful for very cool objects that do not radiate much (or at all) in the visual part of the spectrum. Another popular scheme is the set of *Strömgren* magnitudes, u, v, b, and y, based on measurements made in the ultraviolet, violet, blue, and yellow, with y being very close to V. Any of the apparent magnitudes can be converted to their absolute counterparts via the equation given in the last section. We express these in the three-color UBV system by using the letters as subscripts, e.g. absolute visual magnitude is called M_v, etc.

1.15 Temperature

The colors are a prelude to the study of temperature, and provide us with a quick and convenient way of measuring that quantity. To a reasonably good approximation, stars radiate as if they are *blackbodies*. A body is defined as *black* if it is a perfect absorber of radiation, hence the term. If we expose such a body to radiant energy we might expect its temperature, the measure of internal heat, to rise. In order to maintain a stable temperature, it must then radiate just as much energy as it absorbs, and thus not only does not appear black, but can be quite brilliant.

A blackbody is a very specific case of equilibrium, and its *spectrum*, the way in which the amount of energy that it radiates is distributed with respect to wavelength, is defined by a precise mathematical function that is powerfully sensitive to temperature. As we proceed from long wavelengths to short, or red to blue, we see a slow rise to a peak, followed by a rapid drop (Figure 1.11). The wavelength of the peak emission (λ_{max}) is inversely proportional to temperature: λ_{max} in Ångstrom units equals 2.9×10^7 divided by T in degrees Kelvin, which is called the *Wien Law*. Thus a 3000 K blackbody peaks at 9700 Å, well longward of the range of human vision, whereas one at 30 000 K maximizes at 970 Å, in the middle ultraviolet. Figure 1.11(*b*) also shows the UBV wavelength bands, and now we see how closely color is related to temperature. As the latter rises, relatively more energy is pumped into the shorter-wave bands, and U − B and B − V must become progressively lower and finally negative. Later we will look at the exact relation between B − V and temperature for real stars, whose spectra can actually show fairly significant departures from the blackbody curves.

The total amount of energy produced per square centimeter by the body is expressed through the *Stefan–Boltzmann Law* by a constant, σ (the Stefan–Boltzmann constant), times the temperature raised to the fourth power, T^4. Thus if you double the temperature, the brightness of the surface goes up by a remarkable factor of 2^4 or 16. This effect is illustrated in Figure 1.11(*a*), which shows the amount of radiation emitted by three blackbodies of the same size.

The total luminosity of a spherical body like a star must also depend on its surface area, or number of square centimeters, which is given by $4\pi R^2$ where R is the radius.

a

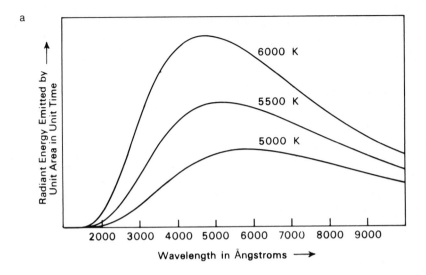

b

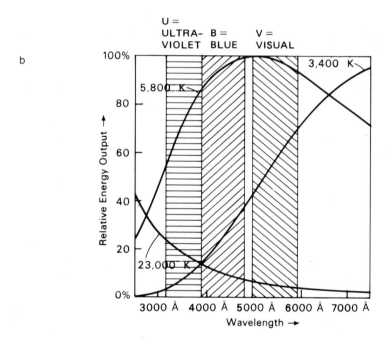

Figure 1.11. The distribution of energy, or intensity, with wavelength for various blackbodies. These graphs are sometimes called *Planck curves*, after the physicist Max Planck who discovered the mathematical function that represents them. (*a*) shows the energy output for three bodies of the same size but of different temperatures. Note that as the temperature increases, the peak of the curve shifts to shorter wavelength and the light becomes bluer, and that the total energy output – represented by the area under the curve – increases dramatically. (*b*) shows three blackbody curves with much greater temperature differences, with the peaks normalized to 100%. The blueward shift with temperature is now much greater. The shaded areas show the U, B, and V bands used for magnitude measurement, and we see how relative magnitudes, and colors (U−B and B−V) will change with temperature. The V magnitude of the cool star will be lower (brighter) than the B magnitude, a situation that will be reversed for the hot star. From *Principles of Astronomy, A Short Version*, 2nd edn., by S. P. Wyatt and J. B. Kaler, Allyn and Bacon, Boston, 1981.

The amount of energy radiated into space, L (in ergs per second) is thus equal to $4\pi R^2 \sigma T^4$. This quantity in turn can be calibrated in terms of absolute magnitude, so that we can connect the empirical astronomical radiation units with the physical.

As mentioned above, however, actual stars do not really behave like the blackbody ideal, perhaps having an excess of energy produced in the red, or a deficiency in the blue. The problem is that a real star emits its radiation from a gaseous, partially transparent atmosphere whose temperature increases with depth. The layers that you observe depend on how transparent the gases are, and their opacity depends upon wavelength. A good analogy is the penetration of red light through a fog that will not allow transmission of blue. Thus the star can appear hotter as we peer deeper at one wavelength, and cooler at another, where we are allowed a glimpse of only the shallower layers. What we actually see is a superposition of many blackbodies that correspond to various layers, each modulated by the effects of the cooler overlying gases: a horribly complex situation.

How then do we relate the blackbody law to what nature actually provides, and what can we mean by the 'temperature' of a star? We mentally substitute for the star a true blackbody of identical radius, and calculate what its temperature would have to be in order to provide the observed luminosity. In practice, we use the above relation, $L = 4\pi R^2 \sigma T^4$, measure L and R, and solve it for T to find what we call the *effective* temperature, which describes the overall nature of the star quite well. The Sun, as an example, has an effective temperature T_{eff}, or T_e, of 5780 K. Its blackbody spectrum falls between the top two curves in Figure 1.11(a).

1.16 Structure

In the previous sections we alluded a bit to the structures of stars; let us now take a more extended look. We divide a star into three broad, somewhat overlapping zones: the core, the envelope, and the atmosphere. There are other layers outside the atmosphere that will be discussed in proper context later.

The action of a star takes place in its core. That is where all the energy is produced through various processes of thermonuclear fusion, in which radiation is created literally out of mass. The principal fusion mechanism, in the Sun and in other main sequence stars, is the formation of one atom of helium out of four of hydrogen. The helium atom is a bit lighter than its four constituents, the lost mass (M) having been converted to energy (E) via Einstein's famous equation, $E = Mc^2$. These *nuclear burning* reactions, as they are frequently called, can occur only within the deep interiors of stars where the temperature is high enough – at *least* 5 000 000 K – to sustain them (which is why it is so difficult to get fusion reactions to take place on Earth for purposes of power generation). Only a fraction of a star's mass is above this critical temperature. In the case of the Sun, the core constitutes only about the inner half of its bulk, which because of the rapid increase of density toward the center means just the inner 25% or so of its radius. The mass of the core increases in a complex way with that of the star, becoming a smaller fractional part as the total mass increases. At the low end, the core constitutes over three quarters of the star's matter, whereas at the higher end it is under 10%.

The *envelope*, which constitutes almost all of the rest of the mass of the star, and from its name obviously surrounds the core, has two related functions. First, in tandem with the nuclear burning zone itself, it provides the gravitational energy needed to drive the core temperature above the critical fusion value. The vast bulk of the envelope squeezes downward on the star's interior, compressing and heating it, much like the action of a piston on the fuel mixture in a diesel engine.

Second, the envelope provides an insulating blanket that keeps the energy generated in the interior from escaping too rapidly, thus keeping the core sufficiently hot. Radiation cannot escape directly to space from the core, but must work its way slowly outward through a process of successive absorptions and re-emissions by atoms (Chapter 2); otherwise the core would quickly cool, snuffing out the fusion reactions. The envelope produces an additional benefit. At the 10^7 K temperature of the solar core most of the energy generated is in the form of γ rays that would destroy all life on Earth. The radiation that falls upon us has actually taken several thousand years to transport itself through this thick overlying layer to the outside. It is then finally emitted to space at the much lower temperature of the surface, generally at, or at least closer to, the more benign optical wavelengths.

The outer layer of the envelope is called the *atmosphere*. This zone has relatively very little mass, but in some ways is the most important to us, as it is the part of the star that we actually *see*. The atmosphere is the region of partially transparent gases from which the energy generated in the core long ago finally escapes; it is the layer in which the *spectrum* is produced, which is the core of this book, and from which we learn almost everything we know of the physical natures of the stars.

1.17 Evolution

Stars are not static, eternal creatures, but in a certain sense are living bodies, a concept that provides a certain anthropomorphic touch to the vocabulary of the subject: we speak of stars being born, of evolving, and of dying. We are way ahead of our story here, since stellar evolution will be examined in considerable depth after we have looked at the huge variety of stars presented to us in the sky. However, it will help to establish a brief view of events, and a context within which to place the various stellar types that we will encounter.

The large majority of stars fall within the 'main sequence', already encountered in Section 1.1, which represents the normal hydrogen-burning stars arrayed according to their masses. (For historical reasons they are equally referred to as 'dwarfs'.) As the mass of a star goes up, so does the size of its core, so that more fuel is available for burning. More importantly, the interior temperature goes up as well. The rate of thermonuclear reactions is exquisitely sensitive to this parameter, with the result that the luminosity of a star climbs *very* steeply with mass: a star with ten times more matter than our Sun is ten thousand times as bright. The more massive stars in fact burn their fuel supplies so rapidly that in spite of the larger amounts available, *they actually survive for considerably shorter periods of time*. Luminous stars have very different characteristics from dim ones, so that the simple existence of a range of masses in itself produces a wide variety of these shining bodies that cover all the spectral classes to be introduced later.

But there is much, much more. The fuel supply is limited, and its exhaustion produces some strange effects. When the nuclear fire expires, the core loses a major part of its support, and it contracts under the influence of gravitational compression and actually heats up, which initiates further fusion on the inner edge of the envelope. The star then actually becomes brighter and the whole envelope balloons outward. It becomes so large (perhaps 50 or 100 times solar) that in spite of the increase in luminosity the atmosphere cools, and we now have a *giant star* (as opposed to the main sequence *dwarf*), a generic name for a broad variety that we will also examine later. As a result of additional nuclear burning (the creation of helium out of carbon for example) and vigorous outflowing winds that cause it to lose considerable mass, the star will undergo several transformations, eventually terminating as a *white dwarf* that is little more than the spent core itself: nearly the entire envelope will have been ejected back into space from which it came.

The most massive, initially brightest stars will behave somewhat differently. As the hydrogen fuel in the core is exhausted these grow even larger than the giants into what are termed *supergiants*, which at their largest are comparable in size to the orbits of the Jovian planets. As evolution proceeds, the cores can become very hot and develop very advanced burning states in which even iron is created. Instead of dying quietly, these stars (and possibly other kinds that are not yet well understood) explode as *supernovae*, producing tremendous blasts easily visible over intergalactic distances. The final products of stellar evolution cause the advancing heavy element content of the Galaxy that produces the different population types we observe.

The variety of stars produced by these evolutionary processes, and others to be described later, is quite astonishing. We see them over an immense range of masses, luminosities, and temperatures; some vary in size and brightness over time scales that range from seconds to years; they appear in a plethora of multiplicities in which the stars can actually interact (and even swallow!) one another as they change in size; and they appear in forms that we frequently cannot explain at all. Our purpose here in this volume is to organize the stars, and to find relations among the different kinds so that we can begin to make some sense of how these cosmic lights work, how they are born, live, and die.

2 *Atoms and spectra*

In order to study the very large we must first examine the very small. We are all, including the stars, made of atoms: to the Greeks the final indivisible part of nature. We know the stars through their radiation, which is produced or altered by atoms and the processes that go on within them. In order to be able to interpret starlight in terms of the stellar characteristics outlined in the previous chapter we must understand exactly how atoms – and their various molecular combinations – interact with radiant energy. Let us begin with a firm foundation, and for the reader not familiar with the physical construction of matter, first examine the natures of these atoms.

2.1 Atoms

The material world is constructed largely of 92 different kinds of natural elements. Several more have been produced in the laboratory, but they do not concern us here. The Greek philosophers were not quite correct. Atoms can be subdivided onto yet more elementary particles, and it is out of different combinations of *electrons, protons,* and *neutrons* that the different elements are assembled. Physicists have discovered a vast array of other kinds of particles that contribute to the matter of the Universe and that are important in understanding atomic processes, but we will not consider these either. The focus of this chapter is to examine the subatomic world only insofar as the formation of stellar spectra is concerned.

The three principal subatomic particles have two simple properties that we need to use: mass and electric charge. Protons and neutrons have similar masses of a mere 2×10^{-24} gram, with electrons weighing in at only 1/1800 that of their heavier relatives. To put these values in perspective, the average human body contains about 10^{29} of these particles, and the Sun – only an average star among the trillions in the Universe – some 10^{57}! Protons and electrons are electrified, and carry equal but opposite charges referred to as 'positive' and 'negative' respectively; like particles repel one another, and unlike attract. The neutron, as attested to by its name, is electrically neutral, and carries no charge at all.

An atom consists of a positively charged *nucleus* that is constructed of the protons and neutrons, which are thereby also called *nucleons*, surrounded by a negatively charged cloud that consists of electrons (Figure 2.1). For convenience we speak of the electrons being in orbit about the nucleus, although by the rules of quantum mechanics

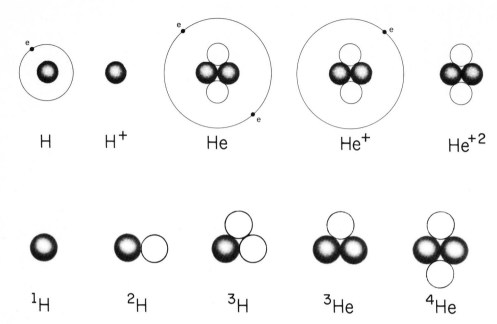

Figure 2.1. The construction of atoms, illustrated by hydrogen and helium. The dark circles represent protons, and the open ones neutrons. Electrons are schematically shown in orbit, and are labeled 'e'. H has one possible positive ion, H$^+$, produced by removing the electron, helium has two, He$^+$ and He^{+2}. The bottom panel shows various isotopes produced by varying the number of neutrons. Adapted from *A Brief View of Astronomy* by J. M. Pasachoff, Saunders College Publishing, New York, 1986.

(the scientific subspecialty of physics that mathematically treats of the atomic world) you really cannot pin down the location of electrons as you would a planet, making the concept inaccurate.

The kind of element is determined solely by the number of protons, called the *atomic number* (A), in the nucleus: hydrogen has one, helium 2, carbon 6, iron 26, uranium 92. Bound to the protons are the neutrons, with a number usually equal to or greater than the proton count. The sum of the two gives the mass of the nucleus and is called the *atomic weight* (M). The atom is represented by a symbol such as $^{14}N_7$, where the upper left superscript is the weight, the lower right subscript the number, and letter the chemical symbol. But since the symbol is tied to the atomic number (i.e. nitrogen, N, has 7 protons), the subscript is usually dropped. In a normal atom, the nucleus would then be surrounded by seven electrons, which are bound to it by the equal and opposite total electric charge (vaguely analogous to the way in which the planets are bound to the Sun) and which render the whole assembly electrically neutral. A complete list of the elements and their properties is given in Table 2.1.

It is illuminating to look at the dimensions of the atoms and of their constituents. In a typical hydrogen atom, the simplest kind, the electron is about 0.5×10^{-8} centimeter (half an Ångstrom unit) from the proton, the only particle in the nucleus. This distance seems tiny until we consider the size of the proton itself, which is only 10^{-13} centimeter across, 100 000 times smaller, so that only one part in 10^{14} of the

Table 2.1. *The chemical elements*

The column headed A gives the atomic number. The column headed M gives the mass number or atomic weight (number of nucleons, protons plus neutrons) of the most abundant isotope of each element, or in the case of radioactive elements, the weight of the most stable isotope.

Name	Symbol	A	M	Name	Symbol	A	M
Hydrogen	H	1	1	Molybdenum	Mo	42	98
Helium	He	2	4	Technetium	Tc	43	99
Lithium	Li	3	7	Ruthenium	Ru	44	102
Beryllium	Be	4	9	Rhodium	Rh	45	103
Boron	B	5	11	Palladium	Pd	46	106
Carbon	C	6	12	Silver	Ag	47	107
Nitrogen	N	7	14	Cadmium	Cd	48	114
Oxygen	O	8	16	Indium	In	49	115
Fluorine	F	9	19	Tin	Sn	50	120
Neon	Ne	10	20	Antimony	Sb	51	121
Sodium	Na	11	23	Tellurium	Te	52	130
Magnesium	Mg	12	24	Iodine	I	53	127
Aluminum	Al	13	27	Xenon	Xe	54	132
Silicon	Si	14	28	Cesium	Cs	55	133
Phosphorus	P	15	31	Barium	Ba	56	138
Sulfur	S	16	32	Lanthanum	La	57	139
Chlorine	Cl	17	35	Cerium	Ce	58	140
Argon	Ar	18	40	Praseodymium	Pr	59	141
Potassium	K	19	39	Neodymium	Nd	60	142
Calcium	Ca	20	40	Promethium	Pm	61	147
Scandium	Sc	21	45	Samarium	Sm	62	152
Titanium	Ti	22	48	Europium	Eu	63	153
Vanadium	V	23	51	Gadolinium	Gd	64	158
Chromium	Cr	24	52	Terbium	Tb	65	159
Manganese	Mn	25	55	Dysprosium	Dy	66	164
Iron	Fe	26	56	Holmium	Ho	67	165
Cobalt	Co	27	59	Erbium	Er	68	166
Nickel	Ni	28	58	Thulium	Tm	69	169
Copper	Cu	29	63	Ytterbium	Yb	70	174
Zinc	Zn	30	64	Lutecium	Lu	71	175
Gallium	Ga	31	69	Hafnium	Hf	72	180
Germanium	Ge	32	74	Tantalum	Ta	73	181
Arsenic	As	33	75	Tungsten	W	74	184
Selenium	Se	34	80	Rhenium	Re	75	187
Bromine	Br	35	79	Osmium	Os	76	192
Krypton	Kr	36	84	Iridium	Ir	77	193
Rubidium	Rb	37	85	Platinum	Pt	78	195
Strontium	Sr	38	88	Gold	Au	79	197
Yttrium	Y	39	89	Mercury	Hg	80	202
Zirconium	Zr	40	90	Thallium	Tl	81	205
Niobium	Nb	41	93	Lead	Pb	82	208

(continued)

Table 2.1. (Continued)

Name	Symbol	A	M	Name	Symbol	A	M
Bismuth	Bi	83	209	Americium	Am	95	243
Polonium	Po	84	210	Curium	Cm	96	247
Astatine	At	85	210	Berkelium	Bk	97	249
Radon	Rn	86	222	Californium	Cf	98	251
Francium	Fr	87	223	Einsteinium	Es	99	254
Radium	Ra	88	226	Mendelevium	Md	101	256
Actinium	Ac	89	227	Nobelium	No	102	254
Thorium	Th	90	232	Lawrencium	Lw	103	257
Protactinium	Pa	91	231	Fermium	Fm	100	253
Uranium	U	92	238	Kurchatovium	Ku	104	—
Neptunium[a]	Np	93	237	Hahnium	Ha	105	—
Plutonium	Pu	94	242				

[a] Elements 93 and beyond do not exist naturally and are produced in the laboratory usually by high energy atomic accelerators.

volume of the atom actually contains mass. We are all mostly empty space! It is the electric charges that prevent us from passing through walls. Obviously there is great room for squeezing matter into small volumes, a theme we will turn to several times in later chapters.

How is it that the nucleus, which contains only positive charges that repel one another, can hold itself together? For an answer we must look to other forces. Nature provides us with four basic forces that are associated with atoms and act over a distance. The weakest of these, oddly enough, is the most familiar: *gravity*, the glue that holds us to Earth, Earth to Sun, and that binds the Universe together. We feel it so strongly only because there are so many atoms in the Earth to pull on us. The next strongest is the *electromagnetic force*, which controls both electric fields and magnetism (two aspects of the same thing: run a current through a wire and you create a magnetic field, pass a wire through a magnetic field and you get a current). This one, of course, is responsible for the charges on the electron and proton. Both of these forces behave according to inverse-square laws: double the distance between two protons, or between you and the center of the Earth, and the force drops by a factor of four.

The third strongest is contrarily called the *weak force*. It is involved with atomic reactions, such as the transformation that can create a proton and an electron from a neutron (called neutron decay: the neutron, a combination of the two particles, is a bit more massive than the sum of the individual masses, and the excess is released as energy). This force is called 'weak' because it contrasts with the other nuclear force, the powerful *strong force*. It is far stronger than electromagnetism but, unlike the electric charge, acts over only *very* short intra-nuclear distances. It overcomes the mutual repulsion of protons only when they are within somewhere around 10^{-13} centimeter of one another, which then allows them to stick together if they are close enough. How they get into such proximity in the first place is a subject for a later chapter. Even so, the strong force of the protons by itself is insufficient to bind the nucleus together; we need neutrons too, which provide additional energy.

2.2 Ions and isotopes

There are two basic variations upon the atomic theme. First, we can change the number of electrons, which produces an atom with a net electric charge and what we call the *ionic state*, or simply an *ion*. The electrons, especially the ones that lie furthest from the nucleus, are not very firmly bound by its electrostatic attraction, and it is fairly easy to strip them away. Atoms in a gas are always moving, with speeds that are proportional to the square root of temperature (which is a measure of heat energy in the gas). Because of their proximity to one another, they are always colliding together, and an especially robust bump can knock an electron (or, with successive collisions, more) off an atom. Or, the atom could be struck by radiation, which can cause an electron to depart.

The nomenclature is simple. If one electron is gone from neon, for example, we call it Ne^+, the plus sign denoting the net positive charge; if two are gone we call it Ne^{+2}, and so on. Neon has 10 protons, so that we could ionize it 10 times (Ne^{+10}) at maximum, leaving us with a bare nucleus. More and more energy – higher and higher temperature – is required to remove successive electrons. It is also possible to ionize an atom negatively. As viewed from close range the electron in hydrogen does not fully cancel the proton's positive charge, and we can attach a second one, giving us H^-, an important component in the atmosphere of our Sun.

The other variation involves changes in the number of neutrons of a given atom, hence in the atomic weight. These different forms of an atom are called *isotopes* (see the lower half of Figure 2.1). For example we can add a neutron to the nucleus of a common hydrogen atom (which is no more than a bare proton) and produce a heavy form called *deuterium*, ^2H (sometimes called 'D'). A second neutron yields ^3H, or *tritium*. These isotopes are *still hydrogen* because only one proton is present.

Most elements actually exist in nature in several isotopic forms, one of which usually dominates: about one percent of all hydrogen is deuterium; helium exists with a tiny amount of ^3He (one neutron is absent) mixed in with the usual ^4He; carbon can be found as ^{12}C, ^{13}C, and ^{14}C, with the first by far the most abundant. Table 2.1 lists the most common forms. However, for each atom there are limits as to how many neutrons can be added or subtracted. Too great a deviation from the norm will cause the nucleus to become unstable, and it will spontaneously break apart into less massive constituents: we then say it is *radioactive*, because it emits radiation as it shatters. There can be no ^2He, with *no* neutrons for example. Most atoms are found with a range of so-called 'stable' isotopes, which will never break up, or decay. Especially at the ends of the range (and sometimes in the middle as well) the isotope will be radioactive but will decay at a slow enough rate that we still find it in nature: ^3H (tritium) and ^{14}C are examples. Many elements, especially the heaviest ones, have no stable isotopes at all, radium and uranium being the best known. The decay of ^{238}U, however, is so long (its *half-life*, the time it takes for a certain amount to decay to half the original mass, is 4.5 billion years) that there are still large quantities in the Earth's crust. However others, notably technetium (no. 43), are so unstable that none is left in our planet; it, and whatever plutonium (no. 94) we may originally have had, have long since decayed into other elements.

When a radioactive atom splits (or *fissions*) it produces a variety of by-products: ^{238}U for example, decays to ^{206}Pb (lead) and several helium nuclei, and ^{232}Th (thorium) to ^{208}Pb. The ratios in a rock of the numbers of daughter products to those of parents, together with laboratory measurements of the half-lives of radioactive atoms, lead to the age of the rock since its crystallization. From samples of rocks in the Earth, Moon, and especially meteorites, we learn that the Earth, the solar system, and consequently our Sun, are 4.5×10^9 years old.

A fissioning atom also releases various kinds of radiation, which is what makes such elements so dangerous to be around. The early classification listed three forms, α, β, and γ rays. We have since found that 'α rays' are actually expelled helium nuclei (which are still frequently referred to as α particles), that β rays are electrons produced through weak force interactions, and that γ rays are high energy electromagnetic radiation.

2.3 Molecules

By sharing their electrons with one another atoms can combine into more complex structures called molecules. Since there is no limit on the number of atoms that can agglomerate, the number of possible molecules is effectively infinite. However, they are relatively fragile, and are easily broken up by collision or high-energy radiation, so that with the higher temperatures of stars all we see are the relatively simple and hardy ones. The cold depths of interstellar space house some of the more complicated varieties.

The simplest molecules are made of just two atoms and are called *diatomic*. Good astronomical examples are CN (carbon plus nitrogen, or cyanogen), CH, OH (hydroxyl), H_2 (two hydrogen atoms), C_2, and the like. Typical three-atom or *triatomic* varieties are CO_2 (carbon dioxide) and H_2O (water vapor). Four-atom molecules are rare in stars, though extended circumstellar envelopes, where the gas may be shielded from stellar radiation by dust, contain these and more.

Molecules can be made of different isotopes, which we designate by atomic weight when it is necessary to do so, such as in differentiating $^{13}C^{14}N$ from $^{12}C^{15}N$. They can also be ionized, such as CH^+, and if the information is required we can combine ionic and isotopic notation, as in $^{13}OH^+$.

2.4 Spectrum lines

Spectroscopy is a science of the relation between radiation and matter, and is at the core of our studies of the stars. Light and its analogues at longer and shorter wavelengths interact with atoms and molecules through their electrons, although atomic nuclei can also play a role, especially at the highest-energy end of the spectrum.

Whatever the lack of precision, and reality, we will continue to use the common notion that electrons orbit their nuclei. The most important principle is that an electron is not confined to a specific orbit. It can have different energies in different orbits, and by jumping around from one path to another (and to yet another) it can absorb or emit radiation. Electron orbits are very unlike their planetary counterparts, however. Whereas the Earth or Mars can take on any size orbit at all (and actually change their distances from the Sun continuously under the influence of the gravitational fields of the other planets) electrons can only move in orbits with very specifically allowed radii that are dependent upon the kind of ion. We say that the orbital diameters, and the energies associated with them, are *quantized* (i.e. only certain quantities are permitted). This property may seem odd at first, but only because it is outside of our normal experience. On the large scale, nature is smooth and continuous; it is quantized only in the range of the very small, and we have to approach atomic proportions and very low energies to be able to see it. The fundamental unit is Planck's constant, h, which we introduced in Section 1.10. Its small size gives us a notion of the scale involved in order that quantization be recognized. The study of such matters is called *quantum mechanics*.

The simplest example of electron orbital structure is provided by hydrogen, which is diagrammed in Figure 2.2. Normally, the electron resides in what we call the *ground state*, in which it is in its smallest possible orbit, where it carries the least allowed energy. The physical radius is about half an Ångstrom: although elliptical orbits can exist that can take the electron closer to the nucleus, the semi-major axis can get no smaller. As we proceed away from the proton the next possible orbit, the second state, also called n (number) $= 2$, is four times (n^2) farther out. This geometric pattern, predicted by the mathematics of quantum mechanics and fully supported by the observations, continues through $n = 3$, 4, and so on, in principle to infinity, but in practice until the orbits become disturbed and cut off by neighboring atoms. In the low-density depths of interstellar space one can detect electron jumps associated with $n > 200$, for which the orbital radius is $200^2(\frac{1}{2}$ Å$)$ or 2×10^{-4} centimeter. The electron cannot find a stable position between any of these orbits.

The electron is tied or bound to the proton via the electromagnetic force: opposite charges attract. To move the electron outward against this field requires energy, which must be given to this tiny particle: it is a potential energy, stored against the time when the electron will drop back downward. The amount of energy required to raise the electron from $n = 1$ (the ground state) to $n = 2$ is very large compared to the increment needed to go from 2 to 3, and each successive orbital increase demands yet less. To go along with the orbital construction, Figure 2.2 also shows such an energy (or *energy-level*) diagram that depicts the values for each orbit. The energy levels proceed upward to a limit, which corresponds to the energy needed to take the electron from $n = 1$ to $n = \infty$, at which point the proton can no longer hold it, the electron goes flying away and the atom is ionized. The limit corresponds to the ionization energy of the atom.

Now, think of an electron in the second orbit or energy level. It may have been raised there by absorption of a photon or through collision with a neighbor. As a general rule of nature, physical configurations always try to adjust to the lowest possible energy,

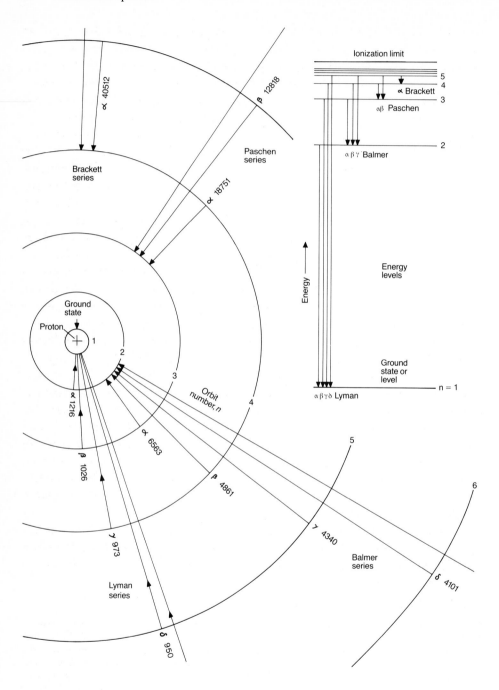

true of atoms and, as we will see, of stars as well. The electron must then drop back down to $n = 1$. It will do so in a very short time, around 10^{-8} seconds, since these outer orbits are very unstable. As it does so it releases its energy as a photon, which goes flying away into space. The energy of this photon must be *exactly* equal to the energy difference between the two orbits (call it E_{21}), which is the same for all hydrogen atoms. Since $E = h\nu = hc/\lambda$, E_{21} strictly defines the frequency or wavelength of the expelled photon. In our example, $\lambda = 1216$ Å, which is in the ultraviolet part of the spectrum.

Next, let us move the electron even farther out, say $n = 4$. It will still seek its lowest level, and it could jump directly from 4 to 1, with the release of a more energetic photon at $\lambda = 973$ Å. However, on the way down the electron could encounter the orbit at $n = 2$, and in stopping there emit energy that corresponds to the energy difference between these two orbits, which corresponds to $\lambda = 4861$ Å, and which is in the blue-green part of the optical spectrum. From there it will jump again to $n = 1$, with the ejection of the 1216 Å photon. Or, it might release *three* photons by going from 4 to 3 (infrared, at 18751 Å), 3 to 2 (red, 6563 Å), and 2 to 1. We have no idea in advance of what any particular electron might do: it is a matter solely of chance. But we can predict what the *odds* are, so that with a large assembly we can calculate *exactly* the number of electrons that might make a given transition per second. The photon emissions that result from the transitions are referred to as *lines* (more properly *emission lines*), as a result of the manner in which they are observed: we will take that subject up later.

The number of possible hydrogen lines is quite startling (think of an electron in $n = 50$), so we must have a way of organizing them. We do so by arranging and naming them in series according to whatever the lower level of the transition might be. Historically the second level is the most important, since the first four downward electron jumps that arrive there radiate in the optical part of the spectrum. All transitions that have $n = 2$ as the lower level are part of the *Balmer series*. The lowest energy of these is $3 \rightarrow 2$, which we designate Balmer α, or more commonly (since the Balmer lines were the first observed) Hα; the $4 \rightarrow 2$ transition is Hβ (at 4861 Å wavelength), $5 \rightarrow 2$ Hγ (4340 Å), $6 \rightarrow 2$ Hδ (4101 Å).

Beyond $n = 10$ or so we simply use the upper number instead of Greek letters, the $15 \rightarrow 2$ jump being called H15, etc. Since the energies of the states get closer and closer together as n increases, so will the energies of the lines. If we plot their positions on a

Figure 2.2. Electron orbits and energy levels, and the various transition series. The left-hand side illustrates the first six possible orbits of the electron as it circulates about the proton, which is designated by a '+' sign. The orbital radii increase by their number (n) squared, that is, orbit 3 is nine times farther from the nucleus than orbit 1 (which has a true diameter of about 1 Ångstrom). On the expanded scale of the diagram, orbit 110, which is responsible for a line in the radio spectrum, would be 85 meters across; in addition the nucleus would have a diameter of but 10^{-5} cm, and could be seen only with a high-powered microscope. The figure shows the first several downward, photon-emitting transitions of the first four series (Lyman, Balmer, Paschen, and Brackett) with their names and wavelengths. The right-hand side shows the energies of the orbits in a vertical scale, also drawn in correct proportion. The ionization limit corresponds to an orbit of infinite radius, which is another way of saying that the electron is then free of the proton, or that the electron's velocity is too great for the electrostatic attraction to bring it back. Drawing by the author.

wavelength scale, we see them pile up against a limit at 3646 Å, which corresponds to the energy required for ionization for an electron already in the second orbit.

The other series are named after their discoverers, and none of the lines that belong to them are optically visible. But they can be photographed or detected by electronic means, and they are all potentially important in astronomy. The ultraviolet lines, whose lower levels are the ground state, and which we used as our first examples, are collectively called the Lyman series, which starts with Lyman α at 1216 Å ($n = 2 \rightarrow 1$) and proceeds to a limit (corresponding to the total ionization energy, called the *ionization potential*) at 912 Å. On the other side of the optical we observe the Paschen series: Paschen α ($n = 4 \rightarrow 3$) begins it at a wavelength of 18 751 Å, and the lines form the limit at an almost-visible $\lambda = 8208$ Å. Farther into the IR are the Brackett, Pfund, and Humphries series that end on levels 4, 5, and 6 respectively. The smaller the energy difference the longer the wavelength, and if we consider transitions between high-lying states the lines fall in the radio spectrum where the series are named after the number of the lower level. The 109 α line ($n = 110 \rightarrow 109$) for example is found at 5007 megahertz (MHz, or million Hertz).

2.5 Absorption lines

The above section dealt exclusively with downward, energy-yielding transitions. The reverse – upward, energy-absorbing electron jumps – are also possible. But whereas an electron will spontaneously jump down to find the ground state, for an absorption to occur we must provide an outside energy source, which can be a supply of photons. Let us take light from a hot blackbody (so that there is enough short-wave energetic radiation) and shine it upon our hydrogen atoms. A blackbody emits what is called *continuous radiation* or just a *continuum*, since photons of *all* wavelengths are present, including those that correspond to the various line energies. If an atom in the ground state encounters a Lyman α photon at 1216 Å (which carries an amount of energy equal to the energy difference between $n = 1$ and $n = 2$), the electron can absorb it and be raised to the second level, thereby depleting the continuum at that particular wavelength.

In order for such an absorption to take place the colliding photon's energy must exactly equal the energy difference between levels. The electron cannot simply absorb part of a photon: it must absorb all or none of it. Consequently, only photons that have energies equal to those energy differences can be so affected. If we have a sufficient number of atoms, they can remove enough Lyman α photons to be noticeable, and upon examining the spectrum, we will see an *absorption line*, and of course we must see the other Lyman lines as well. If the atoms have electrons in the second level, they will absorb photons of the Balmer series, the third level the Paschen series, and so on.

2.6 The formation of spectra

We now look at the real world, and how spectrum lines are actually produced in nature, by the stars and nebulae that abound in space. Let us take a box and fill it with a low density hydrogen gas (low enough so that the atoms do not distort one another's energy levels) as in Figure 2.3. Next shine a source of continuous radiation through it, perhaps from a metal that has been heated to high temperature. We can observe the spectra produced by this system in three ways, by looking at the continuum source alone, by looking at it through the box, and by examining the radiation from the box alone. The device for observing the spectra is called a *spectroscope* or *spectrograph* (depending upon whether one looks visually, or by photography or other permanent recording), which we will examine in detail later: for now just assume that we are able to spread out the light according to its component wavelengths and produce and examine the various spectra.

The spectrum of the light source alone (1 in the figure) will be that of a simple blackbody, a continuum as illustrated by the accompanying graphical spectrum. We covered this matter in Section 1.15. But now let us superimpose the gas in the box upon this continuum. This gas does not radiate like a blackbody at the temperature of the box

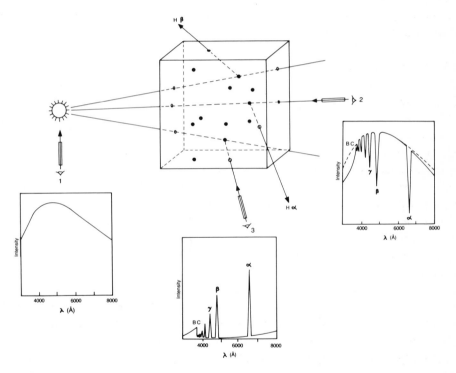

Figure 2.3. The formation of spectra. Light from an incandescent source, perhaps a light bulb, produces the continuous blackbody spectrum observed with spectrograph 1 (here a generic device for spreading the light into its component wavelengths). Photons of all wavelengths enter the box of hydrogen. Those whose energies do not correspond to energy differences between the orbits of H pass through largely unscathed. Those at the wavelengths of the Balmer lines have a good chance of being absorbed, as we see in the second spectrum. The dotted line is the original continuum, and demonstrates that there is also some continuous Balmer absorption (BC), as well as some from beyond the end of the Paschen limit at 8208 Å. The electrons in upper orbits must drop down, producing the Balmer lines and continuum in emission. These can go flying off in any direction and will be seen with spectrograph 3, which is not looking at the continuous source. Diagram by the author.

because the density is too low. The hydrogen atoms are flying around with a great speed that depends upon temperature, and they must constantly collide with one another. In any collision an electron can be knocked from one orbit to another. If it is bumped upward it has absorbed energy of motion from the colliding atom, whose velocity is then reduced, and vice versa. At any given moment the hydrogen atoms will therefore have their electrons distributed among the various energy levels in a statistically predictable manner that depends on temperature and density.

As the continuum radiation passes through the box, the photons whose energies correspond to all the energy differences between level pairs have a chance of being absorbed and eliminated from the spectrum. The number that actually do get removed depends on the number of electrons that have been collisionally raised to the appropriate level, and how long the path length is through the box. What comes out the other end (2 in Figure 2.3) will be an absorption spectrum: a continuum that is broken or depleted at the wavelengths of the hydrogen (in this case Balmer) lines.

Finally, examine the spectrum of the box *alone* (3 in the figure). The energetic electrons in the upper levels, whether raised by collision or absorption of radiation, will always seek to jump downward, and in doing so must radiate the pure emission-line spectrum that we discussed extensively in Section 2.4. To be sure, some of the emitted photons will go flying off in the direction of spectrograph 2. These will act to fill in the absorptions to a small degree, but since the emission must spread out over the whole sphere, there are not enough going in any one direction to make much difference in the absorption lines. In addition, any emission might be reabsorbed by another atom (a process called *self-absorption*), but the net effect will still be an escape of photons in any particular direction.

2.7 Continua

Not only will lines be produced by the gas in the box but so will a continuum, both in absorption and emission. A photon shortward of the Lyman limit at 912 Å has sufficient energy to ionize an electron in the ground state of hydrogen. The energies of the free electrons are not quantized as are those bound to the proton, so that any of these photons can do the job. Our spectrograph 2 in Figure 2.3 must then show a continuous depression of the blackbody spectrum shortward of 912 Å, which is called the Lyman absorption continuum. The probability of such absorption is greatest right at the limit, and will decrease with wavelength. The continuous absorption will thus gradually diminish, and the spectral intensity will again approach that of the original blackbody. We will of course also observe the Balmer continuum (as in the figure) that arises from the absorption of photons shortward of 3646 Å by electrons in $n = 2$; it takes less energy to ionize from the second orbit, since previous collisions or absorptions have done part of the task. The Paschen, Brackett etc. continua must also be present.

We see the complement to this process in spectrograph 3. The free electrons produced by radiative (or collisional) ionization must eventually *recombine* (the process is called simply *recombination*) with some free proton. They can land on any orbit, dictated again by chance, and will now *emit* a photon in the spectral region of one of the above continua. Since the energies of the free electrons are, again, unquantized, the effect will be to produce *emission continua* shortward of 912 Å (Lyman), 3646 Å (Balmer), 8204 Å (Paschen), 14585 Å (Brackett) etc. These will be strongest at the wavelength limits, the continuum *heads*, and will diminish with decreasing wavelength.

In addition, the free electrons (sometimes called the continuum electrons) can change their energies *without* recombination. For example, a fast moving electron may pass close to a proton, not near enough to be recaptured but sufficiently so for it to have its path changed and its speed and energy reduced, with the ejection of a photon. Since there are no restrictions on free orbits (motion of a free electron relative to the interacting proton), the set of all such transitions must be an emission continuum, which will pervade the entire spectrum, especially in the low-energy radio. Conversely, a free electron that is passing near to a proton can accelerate by picking up a photon, providing a mechanism for continuous absorption. What we actually see depends upon the conditions in the gas and the balance between the two mechanisms.

Transitions of the above kind are called *free-free* since the electron is not bound to a proton in either the initial or final states. In the same vein, recaptures are *free–bound* transitions, ionizations are *bound–free*, and ordinary jumps from one orbit to another are *bound–bound*.

2.8 Line structure: pressure, and density

An energy level actually has some width and structure to it. That is, there is a small range of energies the electron can take in any given orbit. Downward transitions between two orbits can then emit over a small range of wavelengths centered on the mean position of the line. Conversely, the upward jumps can extract photons from the incoming radiation field over a small wavelength range. Each line, whether in emission or absorption, will exhibit structure, and will have an associated *profile*, which describes the exact shape of the graph of intensity versus wavelength within the line. We will have a great deal to do with line profiles in the coming chapters as they can tell us much about stars and the constructions of their atmospheres.

The levels always have a certain natural width associated with them that is predictable under the rules of quantum mechanics, but the collisions also play a role, in that a close encounter between atoms can distort the positions of the orbits. Statistically, over the whole assembly of atoms, the effect of collisions is to broaden the lines, that is to allow emission or absorption farther away from the line center, causing an increase in line width.

As the pressure and the incidence and effect of collisions increase, so must the breadths of the lines. If we have a realistic gas, with many atoms and ions that produce spectra with tightly packed and closely spaced emission or absorption features, the lines begin to overlap one another. The gas also becomes so thick that emitted photons are readily reabsorbed and do not escape immediately from the box of Figure 2.3, and every detailed atomic process is balanced by its inverse. This phenomenon, together with the natural occurrence of continua as discussed above means that at some point a gas under high-density conditions (as opposed to the low density assumed for the gas in the box) will produce a pure continuous spectrum. We see then the conversion of a line source into one that perhaps radiates the spectrum (if all the conditions are right) of a blackbody at the temperature of the gas, and we have come full circle from Chapter 1.

We are now in a position to summarize the formation of spectra with three rules known as Kirchhoff's Laws:

1. An incandescent solid or gas under high pressure will produce a continuous spectrum.
2. A low density gas will radiate an emission-line spectrum (with an underlying emission continuum).
3. Continuous radiation viewed through a low density gas will produce an absorption-line spectrum.

2.9 Other atoms and ions

Real astronomical gases, of course, contain a wide variety of atoms other than hydrogen. The spectrum of an atom, or ion, depends upon the number of electrons acting to produce it and the strengths with which they are bound to the nucleus. Since these conditions are different for every species of atom or ion, the spectrum produced by each will be unique: no other atom produces lines that exactly mimic the Balmer series of hydrogen, for example. We can, then, in principle always tell the composition of the gas by what lines are present.

Consider helium, with its two electrons. Normally only one will be excited into an upper state at any one time. If it is in a high energy level the inner electron will screen one of the protons, and the orbits will take on a hydrogen-like appearance. But in a lower excited state the negative charges interact with one another, in effect splitting the analogous hydrogen energy levels, which in turn creates many more possible transitions at wavelengths very different from those of the Balmer (and other) series. There are a half-dozen helium line sequences visible in the optical spectrum alone, with important visible lines at wavelengths of 5876 Å and 4471 Å. In spectral nomenclature, these are called He I λ5876 and He I λ4471, where the Roman numeral 'I' refers to the spectrum of the neutral state and the wavelength (λ) is almost always in Ångstroms (sometimes it is in *micrometers*, μm, 10^{-6} meter or 10 000 Å, or in *nanometers*, nm, 10^{-9} meter, or 10 Å).

If we singly ionize helium and reduce it to one electron, we will get a spectrum similar to that of hydrogen, but because it is held more tightly by two protons rather than one, all the lines will be shifted downward in wavelength by a factor of four (two-squared). Thus the He II (where 'II' refers to the spectrum of the singly ionized state; III then denotes the spectrum of the doubly ionized state, etc.) Balmer series that corresponds to $n = 2$ falls on top of the hydrogen Lyman series, but with one He II line interspersed between each of the H lines. The optical He II lines belong to the higher Paschen and Brackett (or *Pickering* as the latter is usually called) series, the strongest being He II Paschen α at λ4686. Hydrogen and He$^+$ are an example of an *isoelectronic sequence* in which the ions all have the same number of electrons, but for which the spectra are progressively shifted to the UV. This one continues with Li^{+2}, Be^{+3} . . . O^{+7}, etc.

As we increase the atomic number the complexity of the energy level diagrams and of the resulting spectra increases dramatically. Ionized oxygen, O^+, which produces the commonly observed O II spectrum, provides a fine example in Figure 2.4. The energy levels are arranged in close sets called *terms*. Here we see the terms broken into groups of two or four called *doublets* or *quadruplets*. They are arrayed from left to right across the diagram according to a progression of letters S, P, D, F, G, that refer to quantum mechanical properties held in common. For technical reasons the single states that are organized with the doublets on the left, and the single and triple states that appear with the quadruplets on the right are still called doublets and quadruplets respectively. These names refer to more than just the multiplicity of the terms; further explanation requires a considerably deeper examination of quantum mechanics, which is out of place here. Other ions, however, O^{+2} for example, have genuine sets of *singlets* and *triplets*: odd and even term multiplicities alternate with ionic state. Neutral oxygen has *quintuplets*, and more complex ions have terms with even greater multiplicities.

The collections of transitions that connect two terms are called *multiplets*. These line multiplets are named by stating the terms from which they arise. One that connects a doublet in the 'P' column with a doublet in the 'D' column is called $^2P-^2D$. More complex notation can designate individual lines. Note that lines connect only adjacent sets of terms. Other jumps (including doublet to quadruplet) are possible but unlikely and the lines are usually quite weak. The spread or range in wavelength of the individual lines within a multiplet varies considerably from one multiplet and ion to the next. Pairs and trios of very closely spaced lines that arise from doublet or triplet terms are also frequently referred to as doublets or triplets. For example, we speak of the C II $\lambda4267$ doublet, whose two lines are so close together that they cannot ordinarily be separated, and are always blended in astronomical spectra.

The most complicated ions, such as those of iron, have hundreds of catalogued lines. In the spectrum of a mixed gas, all the features of all the ions overlap, falling onto one another in an extraordinary jumble. Absorption lines in a stellar spectrum frequently align to produce *blends*, and sometimes strong ones mask weak ones. In all it can be a difficult task to ferret out the individual ions that produce the observed spectrum, especially for the rare ones that may produce but a few observable features, or even only one. Unless the lines are strong and prominent, identification of an ion in a spectrum is best done by finding all or most of the components of one or more of its multiplets, and by seeing whether or not the line strengths (the amounts of energy extracted from the continuum) match theoretical predictions.

2.10 Molecular spectra

As complex as atomic spectra can be, they are supremely simple compared to their molecular analogues. In an atom the energy levels are caused only by the electrons. But molecules can store energy in additional ways that produce a hierarchy of energy levels. Think of a single level associated with an electron orbit that is produced by the electric

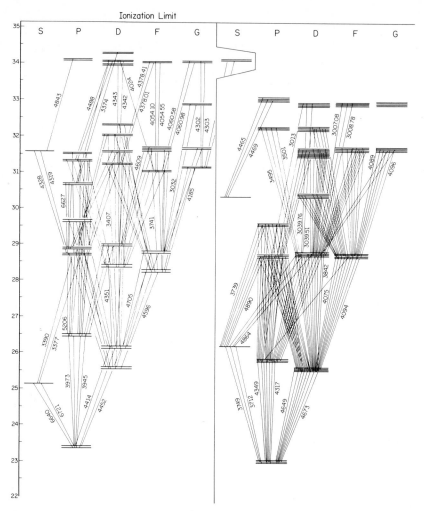

Figure 2.4. Energy levels of the O^+ ion, with the transitions of the optical O II spectrum. The levels for this ion are arranged in groups of one to four called *terms* from which arise *multiplets* of lines that are spread out in wavelength to varying degrees. See the text for a fuller explanation. A chart such as this one is often called a *term* or *Grotrian* diagram. The complexity of the electronic orbital structures of the heavier atoms is awesome. Here we present only the upper part of the diagram that produces the optical transitions. On this scale the ground state is about 40 centimeters off the bottom of the page. Below, we find levels that involve high energy ultraviolet transitions. Most of these terms involve the excitation of the outer (valence) electron only. The horizontal line at the top represents the ionization energy, above which the excited electron is lost to the atom, resulting in O^{+2}. If two electrons can be excited at the same time we can get energy levels above the ionization limit, adding to the complexity of the diagram. Diagram by the author, from *A Mupltiplet Table of Astrophysical Interest* by C. E. Moore, US Govt. Printing Office, 1945.

fields of the electrons and the atomic nuclei of the molecule. First, this level is broken into a series of sub-states caused by the quantized *vibration* of the molecule (in which the atoms oscillate back and forth as if they were on the ends of springs). Then, each of these vibrational states is divided again into sub-sub levels produced by quantized molecular *rotation*. An example is presented in Figure 2.5.

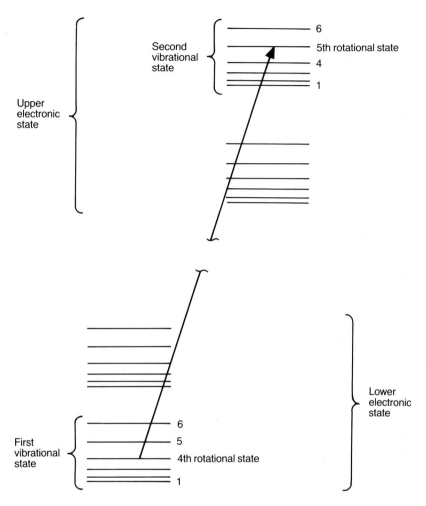

Figure 2.5. Molecular spectra. Two electronic states are shown. Each is divided into vibrational states, of which only the lowest two are drawn. Each of these is split again into rotational states, for which only the lowest six are illustrated. A single molecular absorption line is shown arising from the 4th rotational state of the 1st vibrational state of the lower electronic level, and ending on the 5th rotational state of the 2nd vibrational state of the upper electronic level. The line is a part of *band* of lines created by a set of transitions between the two vibrational states, in which the rotational state number is allowed to change only by plus or minus one. The collection of lines produced between all the vibrational states constitutes a system of bands, all of which replace *one line* in an atomic spectrum. Adapted from *Astrophysics* by L. H. Aller, 2nd edn., Ronald Press Co., New York, 1963.

A single molecular line is produced by a transition between two rotational levels. The collection of transitions between two rotation–vibration sets of states generates a series of lines called a *band* that usually converges either toward the red or blue, depending on the molecule. These bands are the most prominent features of molecular spectra. They are so distinctive that they stand out immediately against a field of atomic lines. The limit, the wavelength at which the rotational lines pile up, called the *band head*, is used to name or categorize the band. Since there are several rotation–vibration sets of sub-levels for each electronic state, a single electron transition will consist of many individual bands. The resulting spectrum of a complex molecular gas, such as the atmosphere of a cool star, can be a nightmare.

2.11 Astronomical spectra

How do all of these descriptions and explanations relate to actual stars? The Sun is definitely not the laboratory demonstration device we posed earlier in Figure 2.3 for the demonstration of the formation of spectra. As we saw in Section 1.16 it is a complex, layered body. However, there are still some analogies to our box of gas. The deep layers are gases under high pressure and they produce a continuous spectrum like the incandescent metal of the figure. As we go outward through the atmosphere of the star the pressure and density drop, and the gas now behaves more like that in the box, which acts to superimpose its complex absorption spectrum upon the outflowing continuous radiation. This outer region is therefore called the *reversing layer*.

If we could see it alone, not against the background of the deep layers, we would observe an emission spectrum analogous to our spectrograph 3 in Figure 2.3. This observation is actually possible during a total eclipse of the Sun. When the moon covers the last of the bright solar surface (known as the photosphere) the absorption spectrum suddenly switches to one in emission called the *flash spectrum* (since it lasts only briefly as the moon quickly covers the reversing layer as well): a dramatic demonstration of Kirchhoff's Laws in action.

Obviously, the notion of a deep layer producing a continuum topped by a thin reversing layer that generates the absorptions is a gross over-simplification. In reality, the pressure drops slowly in the outward direction, and the layers overlap, one merging gradually with another. The continuum and absorption lines are actually created together in the same place; it is just that the lines are on the average formed higher up where the density is lower.

Each line actually is formed at a somewhat different depth in the stellar atmosphere. For example, lines of ions must be produced in layers that are deeper than those that generate neutral features, as it requires higher temperatures to strip electrons from atoms through collisions. Weak lines, for which the photon absorption probability is intrinsically low, tend to be formed in deeper average positions since we must look through a long path-length to see them at all. Strong lines, on the other hand, are related to such a high absorption probability that the gas is very opaque at the wavelengths of their centers: we can then not see very far, and the lines are consequently produced at very high levels in the stellar atmosphere. This situation can actually work to our advantage, as we can then use a variety of lines to study the structure of the atmosphere with depth.

Emission lines are also produced naturally in a variety of astronomical situations; the flash spectrum is but one example. The Orion Nebula, seen in Orion's Sword in Figure 1.1, is a low density gas illuminated from within by a hot star that produces so much ultraviolet radiation shortward of the Lyman ionization limit that it ionizes the nebular atoms. We then see the effect of recombination as the electrons are recaptured by protons and skip down the rungs of the energy level ladder. As another example, interstellar hydrogen produces a radio line at 21 centimeters wavelength caused by a transition that takes place between two sub-levels of the ground state, which is slightly split by a magnetic interaction with the proton. A vast number of emission lines are also radiated in the radio spectrum by molecules – water, carbon dioxide, formaldehyde, alcohols – that are formed in the cold clouds of interstellar space. Some stars have extended low density outer atmospheres that produce emission features that are strong enough to be seen superimposed upon the background absorption spectrum. Others have vast circumstellar envelopes filled with emission-line radiating molecules. The list goes on and on, and we will examine a variety of such cases in later chapters.

One final commentary that is important to the understanding of the stellar spectra that will shortly hold center-stage is necessary. The kinds of atoms and ions that are observed in the spectrum of a star depend obviously upon the composition of the mix, and not so obviously, but *critically,* upon the temperature of the gas. Hydrogen represents a case in point. Some 90% of a stellar atmosphere is composed of this atom. Yet in the coolest stars, we *see no lines*. Hydrogen absorptions are present in the solar spectrum, but the strongest, by far, are those of ionized calcium, Ca^+. In older times, before atomic structure was understood, astronomers thought that the Sun was primarily composed of metals. What is happening here?

The explanation involves the internal energies required to produce the conditions that can in turn produce the lines. In order that the Balmer lines be created in absorption, we must already have H atoms with electrons in the second level. This initial shuffling of electrons around in the orbits is done primarily by atomic collisions (Section 2.6). In the coolest stars, the temperature, and the velocities of the atoms in the gas, are so low that the energies of the collisions are insufficient to bounce the electron upward from $n = 1$ to $n = 2$. In spite of the preponderance of hydrogen then, no optical Balmer lines can be observed.

In the Sun, the temperature is high enough that the Balmer lines can be seen. But even there, only one H atom out of 10^8 is actually in $n = 2$ at any given time, the low population again due to the relatively high energy jump from the first to the second level relative to the energy of the gas. The Ca II lines at 3933 Å and 3968 Å, however, arise from the ground state (such features are called *resonance lines*), which contains almost *all* of the Ca^+ electrons. Consequently, even though hydrogen is some 10^5 times more abundant than calcium, the Ca^+ lines are considerably stronger. The moral is that we cannot just look at the kinds of spectra present. In order to interpret the strengths of the absorption lines properly we must also factor in the excitation conditions. This matter is the crux of spectral analysis, and it is of critical importance in our categorization of stellar spectra, to be elucidated in the next chapter.

2.12 The Doppler Effect

Spectrum lines can be identified exactly by their wavelengths, but only if the source and the observer are at rest with respect to one another. If they happen to be moving toward or away from one another, the line positions will be shifted in proportion to the relative speed along the line of sight, the so-called *radial velocity* (see Section 1.9).

Think of a stream of waves coming at you from a stationary source for which you make measurements of wavelength and/or frequency. Now approach the source at a steady velocity. You must encounter the waves more often, giving you the perception of a higher frequency and concomitantly shorter wavelength. That is, an observed spectrum line will be shifted toward the blue part of the spectrum. The faster you move, the greater the shift. It does not matter whether you approach the source, or it approaches you, the effect is the same. Conversely, if there is a relative speed of recession, the line will be displaced redward.

This phenomenon is known as the *Doppler effect*. It is most familiar in terms of sound waves, which makes the pitch of an approaching automobile horn, or airplane noise, appear higher as it approaches, and lower as it recedes. The Doppler Effect is quantified by saying that the percentage wavelength shift is proportional to the ratio of the velocity to the speed of light. Or, if we let $\Delta\lambda$ be the displacement, observed wavelength (λ_o) minus true wavelength (λ_r), then $\Delta\lambda/\lambda_r = v/c$, where v and c are the source and light velocities. If we know the true wavelengths of spectrum lines from laboratory measurements made with everything at rest, and can measure the actual wavelengths (which we do with spectrographs: Section 2.14), we can determine the radial velocity, and so provide data on how the stars and galaxies are moving through space.

Stellar Doppler shifts are small, usually an Ångstrom or two at the most, so that in practice the displacement does not greatly confuse line identifications. We must merely find the velocity that brings all the observed lines into accord with laboratory measurements. Doppler shifts for the rapidly receding galaxies can be very large however, tens and hundreds of Ångstrom units. For the distant quasars they are so great that the UV below Lyman α can actually be seen in the optical. For these the simple Doppler formula for speed no longer works, and we must apply rules developed from the theory of relativity.

It is the Doppler effect that allows us to detect binary stars that are so close together that the components cannot be individually resolved at the telescope under any magnification (see Section 1.6). Unless the plane of the orbit is exactly perpendicular to our line of sight, each star will alternately have a component of orbital motion first toward us then away as it wheels around the other. The result is a set of spectrum lines (or two sets, if the stars are of comparable brightness) whose wavelengths periodically shift back and forth. From the degree of displacement we can infer the orbital characteristics of the binary and use them to derive information on stellar masses.

2.13 The dispersion of light

We have discussed spectra in great depth without examining the devices that are used to produce them, and without actually demonstrating what a real spectrogram (the recording of the spectrum) looks like.

To produce a spectrum we must *disperse* the light from a source, and separate it into its component colors. That can be done in two ways. The original method was to use a glass prism as in Figure 2.6. The shorter-wavelength components of the incoming white light are refracted more both upon entry and exit, and are consequently bent through a greater angle than the longer wavelengths. Thus we achieve a spectrum. Hold a prism up to the sunlight and cast its spectrum on a wall to see for yourself. A rainbow is made the same way by the refraction and dispersion of sunlight through raindrops.

The second and more modern way is to use a *grating*, which employs the principle of *diffraction*. Think of two slits, or apertures, cut into an opaque plate, which is illuminated in Figure 2.7 by a single-color (*monochromatic*) light from the left. Each slit behaves as a new light source, sending radiation to the surface at the right, which they illuminate. The center position, labelled '0', is equidistant from the two slits so that the light waves fall on top of one another and *constructively interfere*; that is, they add together to produce a bright spot of light. A little bit away from either side of this point however, we can find a zone on which the path length to one slit is half a wavelength longer than it is to the other. Consequently, the waves *cancel* one another by *destructive* interference, and the position appears dark. A little farther away, however, (at point '1') the path length difference is *one* wavelength and we have constructive interference, and a bright spot once again. As we proceed farther and farther away, every time the path length difference equals an integral number of wavelengths, we will find a bright *fringe*, as it is termed; the set of these are called *interference fringes*.

Now consider the slit pair to be illuminated by *white* light. Because red rays are longer, we have to move slightly farther away from the center spot to find the first fringe than we do for the blue rays. The effect is that all the fringes (except the one in the center) appear colored, with red at the edge more distant from the center, and violet at

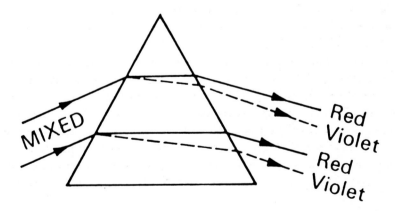

Figure 2.6. The dispersion of light by a prism. Incoming white light is dispersed by refraction into red and violet components, with all the other colors in between. The beam of white light is wide, and the colors produced by the individual rays overlap, producing a washed-out spectrum of low purity. The color discrimination can be markedly improved by restricting the width of the incoming beam with a small aperture or slit.

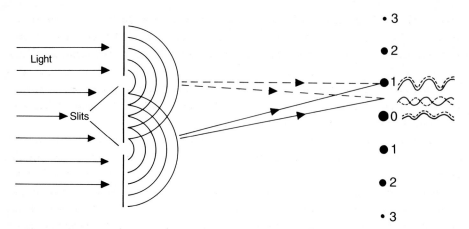

Figure 2.7. Diffraction. Monochromatic (single wavelength) light illuminates a plate with two slits cut into it, each of which acts as a new light source. The radiation from the apertures falls upon a distant surface in a series of interference fringes. (The proportions are not drawn to scale). At the center position ('0'), the distance to the two slits is the same and the waves from each fall on top of one another, *constructively interfering*, to produce a bright spot. Halfway between the '0' and '1' positions, the difference in the distances to the slits is one-half wavelength, so that they fall out of phase, canceling each other by *destructive interference*, and producing a dark zone. At positions '1', called the *first orders*, the difference in distances is one full wavelength, again producing constructive interference. At each order (2, 3, etc.), the waves are offset by another full wavelength. The result is an interference pattern with successively fainter bright *fringes*, interspersed by dark ones. If the incoming light is white, the fringes will be colored. The longer the wavelength the farther from '0' will be the position for which the waves from each aperture will overlap, so that the farther edge of each fringe will be red, the nearer violet. A large number of slits will produce a series of beautiful spectra. Diagram by the author.

the other. We have produced a crude spectrum; actually we have created a large number of them, one for each fringe.

If we add a third slit we can define the color separation somewhat better; if we use hundreds or thousands, we can generate superb spectra. The devices that perform this function are called *gratings*. They consist of a glass plate upon which we rule thousands of narrow lines, hundreds per millimeter. We can replicate one by pressing it onto a plastic sheet, which when coated with aluminum, becomes a *reflection grating*. The different spectra corresponding to each fringe are called *orders*, numbered as in the diagram. The higher orders produce the more widely spread-out spectra.

You can easily see diffraction spectra for yourself by holding a compact audio disk (which has closely spaced grooves that form a reflection grating) up to the light. In a natural setting the colored rings that you see against the moon on a night with filmy clouds overhead are diffraction spectra caused by moonlight passing through 'slits' produced by the spaces between water droplets or ice crystals.

Before comparing these two methods we must define important concepts of spectroscopy known as *dispersion* and *purity*. The dispersion of a spectrum is the degree to which the various wavelengths are spread out. We can indicate it in terms of the wavelength band, in Ångstroms, per degree of refractive angle. Astronomers actually use linear measurement, with the spectrum projected or focused upon a surface. Then we measure the dispersion in terms of Ångstroms per millimeter, or Å/mm. As a practical example again just use a prismatic solar spectrum projected on the wall – measure its width in millimeters, and divide that number into 4000 Å, the optical span from violet to red, to find the average dispersion.

However, you may notice that in your simple solar spectrum that you see no absorption lines, even if your spectrum appears to be very long. That is because the *purity* is bad. The colors mix together too much and overlap to the degree that the lines are filled in and invisible. The Sun is an extended body and the radiation from one edge of it produces a spectrum that is shifted relative to that from the other edge. We see this kind of mixing effect produced by the broad beam of light entering the prism in Figure 2.6. As a consequence, the combined spectrum is smeared to the point where no lines can be seen. In order to define the colors better, so that each position in the spectrum corresponds better to a single wavelength, rather than to a mix of them, we must better define the entry angle of the light. We can greatly improve the spectral purity by first passing the light through a narrow slit or small aperture, which you can make by setting the edges of two razor blades very close together.

With this knowledge at hand let us look at prisms versus gratings. There are serious problems with the former. The dispersion of a prism is not linear, that is, the numerical value in Å/mm changes drastically with wavelength, and is much greater in the blue than in the red where the colors become jammed together. The dispersions of grating spectra, however, are very nearly linear: the dispersion does not change with wavelength. They are excellent devices for examining long-wave spectra, and for the precise measurement of wavelength for the purposes of line identification and determination of radial velocity. Next, it is difficult to increase the dispersion of a prism: we have to use multiple prisms, which absorb considerable light. With gratings we can simply substitute one that has more lines per millimeter, or go to a higher order, both of which increase the dispersion and the purity. In addition, prisms are transmission devices, and glass is opaque to ultraviolet; we must use special glasses, or quartz. A grating can be (and usually is) used in a reflection mode, avoiding the problem altogether.

One major disadvantage of a grating is that it wastes light by dispersing it into many orders that go unused. We can alleviate this difficulty by tilting, or *blazing*, the reflecting planes of a reflection grating in order to throw the majority of the radiation into one pre-selected order. When we consider all the qualities of the two dispersing modes, blazed gratings are the clear favorite.

2.14 The slit spectrograph

Now, how do we put all this together in practice? We must make a device that will accept starlight, transfer it to a dispersing device, and focus it onto a detector. The astronomical spectrograph, the machine that performs the task, is diagrammed in Figure 2.8.

The light from a star first comes to a focus at the focal plane, which is where the spectrograph takes over. The image of a star is never a point: it is smeared by atmospheric seeing (twinkling) and the telescope optics into a disk that may be large enough to affect the spectral purity. Consequently, we isolate a narrow section of the light with a *slit* placed at the focal plane that is the entrance to the spectrograph. After passing through the slit the light now diverges. Before dispersal the rays must again be rendered parallel, as they were as they came to us from the star. Otherwise, they enter the prism or grating at different angles, with the destruction of purity. This parallelism is produced by a lens, or preferably a mirror (to avoid ultraviolet absorption) called the *collimator*.

Now the rays fall onto the reflection grating (or through the prism). Each narrow ray will be broken up into its component colors. If we allowed the whole beam to fall onto a surface, the spectra from the original rays would all overlap, and we would see a jumble with another degradation of purity. We must then use a *camera lens* or mirror to focus all the individual rays of the same color to the same point. Then, if we put a surface at the camera's focal plane we would see a clear, brilliantly colored spectrum, with all the absorption (or emission) lines prominently displayed.

We could remove this surface and use an eyepiece to examine the focal plane, and we then would have a *spectroscope*. If we placed a photographic plate or piece of film there, to be exposed and developed later, we would have a *spectrograph*. The result is a *spectrogram*. These are almost always done in black and white on photographic emulsions especially designed to be sensitive at low light intensities.

However, a small problem arises now. The image of a star on the slit is very tiny. Consequently the *width* of the spectrogram, the dimension perpendicular to that in which the light is dispersed, must be very small, too small perhaps for proper examination and measurement. In order to increase the visibility of the lines, we must *widen* the spectrum, which we can easily do by moving, or trailing, the image of the star along the slit, which has deliberately been set exactly perpendicular to the dispersion. An example of the final product appears in Figure 2.9. *Now*, we see why spectral absorptions and emissions can be called *lines*; they look like narrow lines set across the direction of the spread of colors. Such widening of course is unnecessary when taking spectrograms of extended objects such as the Sun or nebulae: the breadth of the object produces its own widening.

Measurements of wavelength are done by recording a *comparison spectrum* of an element whose wavelengths are precisely known alongside that of the star. For example we might build an iron-arc device into the instrument. We place two bars of iron close together and pass an electric current from one to the other, which produces a hot, electronically excited iron gas in the space between. The light from this gas, which has a pure emission-line spectrum, is passed through the slit with the aid of mirrors, and is made to fall on either side of the stellar spectrum as we see in Figure 2.9. Since we know the iron wavelengths, we can calibrate λ against linear position, and thereby precisely

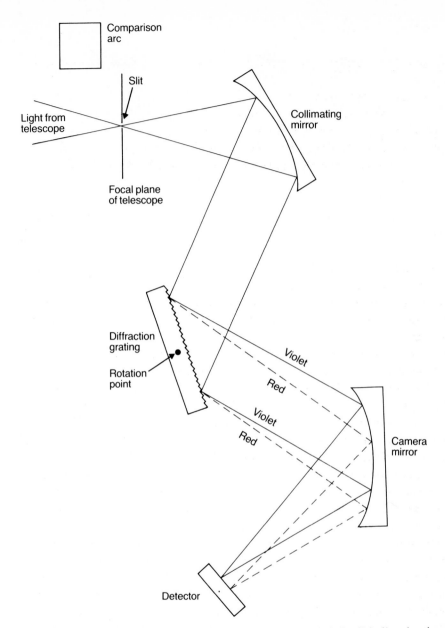

Figure 2.8. A schematic (not to scale) diagram of a modern slit spectrograph. Starlight from the telescope is focused onto a narrow *slit* (or other small aperture), which extracts only a small segment of the image, defining a point source of light for the spectrograph. The diverging cone of radiation is then rendered into parallel rays by the *collimating mirror* (or *collimator*), which sends the light to the *grating* for dispersion. The parallel rays of the same color are then focused by the *camera mirror* onto a *detector*, which may be a photographic plate, or an electronic device. In this diagram, the spectrum would appear as a narrow line with a width perpendicular to the dispersion appropriate to the size of the stellar image. For photographic representation it can be widened by running the star along the slit, perpendicular to the plane of the paper. A comparison spectrum of iron, neon, or other element can be impressed on the spectrogram (or used to calibrate electronically recorded spectra) by passing light from the box labeled 'comparison arc' on to the slit by means of a mirror. The grating can be rotated in order to bring different parts of the spectrum onto the detector in case the dispersion is so large that not all of it will fit. Diagram by the author.

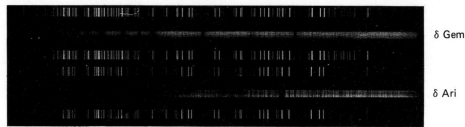

Figure 2.9. The blue photographic spectra of two stars, δ Geminorum and δ Arietis, both of which have rich absorption spectra. Spectra of the comparison arc flank each stellar spectrum. From the known wavelengths of the iron lines, which have been measured in the laboratory, we can precisely measure those of the stellar absorptions for purposes of identification and the measurement of radial velocity. Although the stars have about the same chemical compositions, the spectra appear totally different, an effect solely of temperature. The spectrum of δ Gem shows the progression of Balmer lines, which get closer and closer together toward the blue. In that of δ Ari note two strong lines just to the left of center caused by ionized calcium, and further left a wide absorption produced by the CN molecule. These differences will be a major subject of exploration in later chapters. University of Illinois Prairie Observatory spectrograms, courtesy of K. M. Yoss.

determine the wavelengths of the stellar (or nebular) lines for purposes of identification and radial velocity measurement. We commonly use helium–neon or argon gas mixtures as well, which we excite by passing an electric current through a gas-filled tube.

The photographic plate, while providing us with a wonderful wavelength range and a high information density, is very inefficient and slow. We have to expose for long periods – hours – to bring up spectra of even fairly bright stars, depending of course on the size of the telescope. The recent past has seen the development of far more sensitive electronic detectors, which record the spectral intensity data digitally on magnetic disk or tape for later analysis and graphical display. The charge-coupled device (CCD; see Section 1.11) is becoming the most common. Such an instrument is often coupled with an *image tube*, which electronically amplifies the weak spectral signal before it falls upon the detector. These spectra too can be calibrated by comparison arcs. Widening of the spectrum is unnecessary, so we substitute a small aperture for the slit. An example of the resulting data, made with a television-like scanning device, is shown in Figure 2.10. Obviously, a space telescope *must* record a spectrum in this fashion for relay to the ground via radio telemetry.

A real, working astronomical spectrograph is a complex and versatile device; one such is shown in Figure 2.11. It comes with a variety of gratings, each with different numbers of lines per millimeter and different blaze angles so as to select particular orders. We can then choose a dispersion that is appropriate to the research problem at hand. The greater the dispersion the longer the exposure must be; to reach the faintest stars in reasonable time, the dispersion must be reduced. We might select a low dispersion of say 250 Å/mm if we merely wish to classify a star; or a higher one with 40 Å/mm if we want more quantitative information on line strengths and structures. We

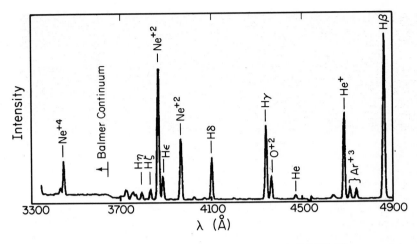

Figure 2.10. A digitally recorded spectrogram of a planetary nebula called IC 351. The spectrum is scanned electronically by a video-like device that records the intensity of a certain segment of wavelengths on magnetic tape for later playback-and graphical display. The nebula is a low density cloud of gas surrounding a hot, illuminating star, and therefore produces an emission-line spectrum. Note the progression of Balmer lines Hβ through Hη, as they merge with the Balmer continuum at 3646 Å. Lines produced by a variety of other ions are also shown. Author's spectrogram from the National Optical Astronomy (Kitt Peak) Observatories.

can also rotate the grating so as to place different spectral regions, from ultraviolet to infrared, onto the surface of the detector.

Spectrographs that work at these moderate to low dispersions (like the one in the figure) generally can be attached directly to the telescope. One placed at the Cassegrain focus, in which the light is brought to a point behind the telescope objective, is a *Casssegrain spectrograph*. If one wants very high dispersion it is usually necessary to go to huge fixed systems that are placed at the *coudé focus*. The coudé focus is a position in the observatory building where the light is brought by a system of mirrors, and which remains fixed as the telescope is moved. These *coudé spectrographs* are capable of enormous dispersions of the order of one Å/mm, allowing us to see extraordinary detail within a line. Solar spectrographs can go to yet higher dispersions, since there is so much light available.

2.15 The slitless spectrograph

It is possible, and very useful, to combine the optics of a telescope and spectrograph into a slitless version of the latter. We merely place a prism in front of the objective the telescope, so that the light is dispersed *first*. Any star in the field of view will then produce a spectrum on a photographic plate placed at the focus of the telescope. Such a device is also called an *objective prism spectrograph*.

The spectral purity is naturally limited since we rely on the stars to be point

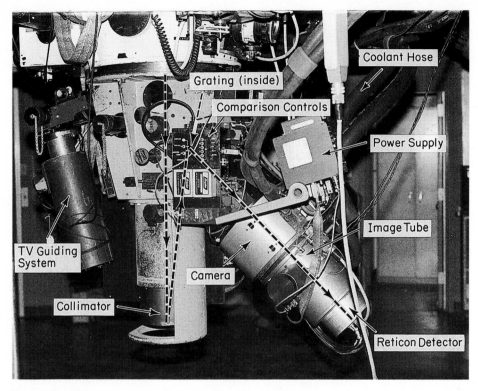

Figure 2.11. An astronomical spectrograph. The various parts diagrammed in Fig 2.8 are labeled on the spectrograph used at the Cassegrain focus of the University of Arizona Steward Observatory's 2.3 meter telescope. The dashed line shows the path taken by the incoming starlight. The 'Reticon' is yet another form of electronic detector, one that consists of an array of light-sensitive diodes that converts photons into an electric current. The light is first amplified before it encounters the reticon by an electronic 'image intensifier'. Photograph by the author.

sources (which they are not), so that the dispersions are generally kept low. The prism is then usually in the form of a wedge with an angle measured at the apex of only a few degrees in order to keep the spectra fairly short. In addition, lower dispersions are necessary to avoid overlap in crowded fields of stars.

Slitless spectroscopy is most commonly employed in conjunction with wide angle Schmidt cameras (telescopes that combine a refracting correcting plate with a focusing mirror in order to achieve wide sky coverage). It is a survey technique used to obtain large numbers of spectra all at one time. As such it is eminently suited for quick classification of large numbers of stars into their various types. However, calibration spectra cannot readily be impressed, so it is quite difficult to use the device for wavelength measurement, which must be done with the slit spectrograph. That device, though, can ordinarily be used to observe only individual stars, so that a survey conducted with one would take an inordinately long time. In practice, we look for interesting objects at low slitless dispersions, then attack them at high dispersions with the more versatile slit instrument. An example of an objective prism spectrogram is shown at the beginning of the next chapter, in Figure 3.1.

Now we are prepared to begin. Let us look at the stars.

3 *The spectral sequence*

After these broad introductions we again turn to the array of letters – OBAFGKM – with which we opened this book. In the last chapter we explored from the perspective of modern astrophysics the way in which spectra are produced by stars. But this kind of analysis is distinct from what we wish to examine now. The physical origin of the spectrum was not finally understood until the 1920s, when the equations relating ionization and excitation to temperature and density were derived, whereas stellar spectra were being observed a century earlier. A common first step in the comprehension of a body of knowledge, be it astronomy or zoology, is the classification of the data. Without order, the causes of the observed phenomena may remain forever hidden. So now we look at the organization of the stars, of the grand array of spectra we see in Figure 3.1. As we proceed through our discussions we will gradually incorporate the origins of the observed phenomena, as we develop the theoretical framework that rests on, yet does not disturb, the foundation of the classification.

3.1 The Fraunhofer spectrum

The spectrum of the brightest star, the Sun, was first mapped by Joseph von Fraunhofer in the early 1800s. He observed over 500 absorptions and assigned letters to the most prominent of them, proceeding from red to blue, long before any chemical origin could be known (Table 3.1). The letters have no relation to chemical symbols nor to those denoting the modern spectral classes of stars. The Fraunhofer designations, however, still provide a convenient shorthand for us today. For example we speak of the *H and K lines*, which are the strongest in the solar spectrum and which are produced by singly ionized calcium: Ca II λ3968 and Ca II λ3934. Three other Fraunhofer lines, C, F, and h, are created by Hα, Hβ, and Hδ, members of the famous hydrogen Balmer series (Section 2.4) that continues on into the ultraviolet. The Hγ line was missed as it lies just redward of the molecular G band, and Hε is lost in the middle of Fraunhofer H. Three of the red lines are caused by terrestrial oxygen and water vapor superimposed upon the solar spectrum as a result of the passage of sunlight through our atmosphere.

Fraunhofer also undertook the observation of stellar spectra and by the 1820s had examined a variety of the first magnitude stars with an objective prism spectroscope. Even with this relatively primitive equipment he was able to see the differences among

Figure 3.1. Objective prism spectrogram (Section 2.15) of a star field in the Milky Way in Taurus. A prism is placed over the telescope objective, in this case the Michigan 24-inch Curtis Schmidt, which produces a spectrum of every star in the field of view, permitting rapid classification. Note the variety of types: some stars feature the Balmer lines of hydrogen, others H and K of ionized calcium. The Henry Draper spectral catalogue (Section 3.3) was compiled from similar plates. The Observatory of the University of Michigan, courtesy of K. M. Yoss.

them, noting the strong green and blue lines in Sirius' spectrum, the similarity of Pollux's with the Sun's and the separate character of that of Betelgeuse.

By the mid-1800s the increasing power of telescopes and spectroscopes both enabled and required astronomers to search for an effective classification scheme to bring some order to the increasing complexity of the observations. These efforts culminated in 1901 with the *spectral sequence*: the *Harvard*, or *Draper*, system of letters that is our principal subject, and which was of such simplicity and versatility that it was able to evolve into the more comprehensive notation that we use today. The beauty of the method was that stars could and still can be placed in order without any reference at all to physical quantities. This independence of observation and theory not

Table 3.1. *The Fraunhofer lines*

The middle column gives the wavelength of the line in Ångstroms. The first three, in the red, are formed by the passage of sunlight through the Earth's atmosphere; the last is in the deep violet.

Name	λ (Å)	Origin
A	7594	terrestrial oxygen (O_2)
a	7165	terrestrial water vapor
B	6867	terrestrial oxygen
C	6563	Hα
D	5890, 5896	neutral sodium (Na I)
E	5270	neutral iron (Fe I)
b	5167, 5173, 5184	neutral magnesium (Mg I)
F	4861	Hβ
d*	4384	neutral iron (Fe I)
G	4300	CH band
g*	4227	neutral calcium (Ca I)
h*	4102	Hδ
H	3968	ionized calcium (Ca II)
K*	3934	ionized calcium (Ca II)

* Not an original Fraunhofer designation: added later.

only allows us to move the science forward before the theory is developed, but is intrinsic to its creation and elaboration.

3.2 Beginnings: Father Secchi

A cursory survey of stellar spectra at first shows what appears to be a bewildering variety. But closer examination reveals that most fall into only a few relatively distinct groupings that correlate with stellar color: some blue-white stars have powerful lines of hydrogen, while others have helium; there is a broad collection of yellow-orange stars with numerous lines of neutral and ionized metals; and we see the set of red stars to be dominated by the banded spectra of molecules. A yet more penetrating view discloses that the groups, however they may be defined, merge smoothly into one another, which demonstrates an underlying unity in stellar nature, now known to be a general similarity in chemical composition. While there may be great and often fascinating variations in the details, recall from Section 1.5 that hydrogen is almost always the dominant element, followed by helium, then oxygen, carbon and others. As a common rule, the greater the masses of the atoms, the fewer there are of them.

Of the many early attempts at classification, one developed by Father Angelo Secchi during and following the 1860s stands out, both for its simplicity and for its foreshadowing of the modern scheme. Readers of older literature will still commonly find references to it. After a number of preliminary variations, he settled on a system of five basic types that closely parallel the descriptive groupings above:

type I   – strong hydrogen lines: blue-white stars like Sirius and Vega;

type II   – numerous metallic lines (sodium, calcium, iron), weakened hydrogen: yellow or orange stars such as the Sun, Capella, Arcturus;

type III   – prominent bands of lines, each of which gets darker toward the blue, also metallic lines of type II: orange to red stars like Betelgeuse and Antares;

type IV   – bands that shade in the other direction, deep red stars of at least magnitude 5: few visible to naked eye;

type V   – *bright* spectrum lines, either in conjunction with or instead of absorption lines: rare.

The first three types are illustrated in Figure 3.2, which displays a variety of turn-of-the-century spectrograms employed in the construction of the Harvard classification scheme (Section 3.3).

The bands of type III were discovered in 1904 to be caused by titanium oxide. Secchi himself found the origin of the type IV bands to be compounds of carbon. The bright, or emission, lines of type V indicate the existence of a hot gas under low pressure, and we now know they frequently imply an extensive, rarefied atmosphere around a star (often produced by processes that cause matter to be lost from its surface). Diffuse and planetary nebulae such as the *Orion Nebula* and the *Dumbbell* also display bright emission lines (see Figure 2.10). E. C. Pickering of Harvard College Observatory later modified the definition of the 'fifth type,' as we frequently see it referred to, restricting it to hot stars with prominent emission lines of helium, carbon, and nitrogen (the *Wolf–Rayet* stars), and to the planetary nebulae. The Secchi types are roughly compatible with the divisions we can discern with the naked eye from the stars' colors alone.

3.3 The Harvard system

With rapidly improving instrumentation, it quickly became clear that the Secchi types were too broad: any single class contained a considerable variation among the stellar spectra. After a half-dozen attempts at more sophisticated classification by others, Pickering enlarged upon the Secchi system with a new sequence based on capital Roman letters. Thus was begun a vast work, all based on photographic spectra. It was financed from the estate of Henry Draper, the first person to photograph stellar absorption lines. All the spectral catalogues produced at Harvard at the time were published in the *Harvard Annals*, and are part of the *Henry Draper Memorial*; stars are still extensively called by their 'HD numbers', as taken from the immense final version.

The first catalogues, which use A through Q in alphabetic progression (Table 3.2) were published in 1890 in *Annals* volume 27. The classification itself was done by Mrs. W. P. Fleming, the first of a famous trio who were instrumental in developing the work. Pickering had doubts about the reality of C, and about the differences between E and G and among H, I, and K. In a study of southern clusters published as pt. 2 of vol. 26 six years later, he and Mrs. Fleming dropped D, L, and I, made K more like G, and

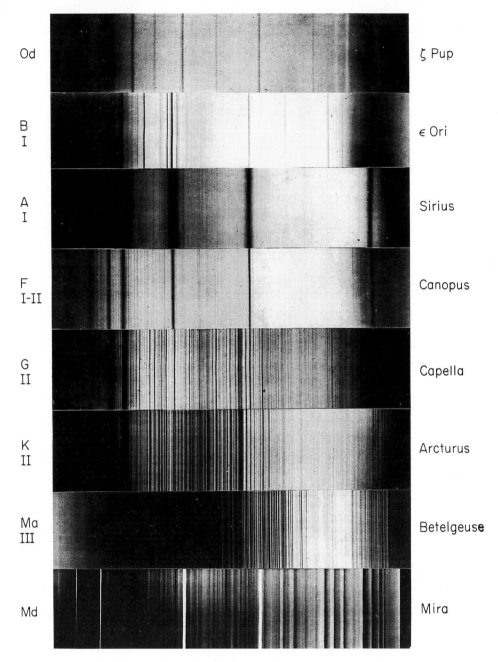

Figure 3.2.

Table 3.2. *The original Draper classes*

Secchi		Draper (Harvard)
I	A	strong, broad hydrogen lines
	B	like class A, but with the addition of the 'Orion lines' (found in many stars in Orion, later determined to be neutral helium)
	C	doubled hydrogen lines
	D	emission lines present
II	E	Fraunhofer 'H' and 'K', and the Hβ lines are seen
	F	similar to class E, but with all the hydrogen lines present
	G	the same as F but with additional lines
	H	the same as F but with a drop in intensity in the blue part of the spectrum
	I	like H except with additional lines
	K	bands visible in spectrum
	L	peculiar variations of K
III	M	Secchi's third type
IV	N	Secchi's fourth type
Pickering V	O	spectra with mainly bright lines (Wolf–Rayet)
	P	planetary nebulae
	Q	all other spectra (changed to designate novae in 1922)

Figure 3.2. Original classification spectra. The panel shows a variety of spectra that illustrate Secchi classification, made with an objective prism and an 11-inch telescope in the 1890s. The stars are arranged downward in order of class, which is also in order of decreasing temperature. The progression of lines and the continuity of types is obvious. The quality of such century-old spectra, from which the modern or Harvard classification system was developed, is quite remarkable. The Harvard classes originally applied to each star, which have since been modified, are given at the left of each spectrum, with the Secchi types immediately below. Note the strong hydrogen lines in Sirius' spectrum, and the powerful H and K Ca II doublet that develops at shorter wavelengths (to the left) for Canopus, Cappella, and Arcturus. Betelgeuse shows a rich spectrum of neutral lines and a suggestion of a molecular band at the far right: the banded structure (caused mostly by titanium oxide) of the cool star Mira is quite dramatic. Mira, a variable star, also displays hydrogen emission lines that are characteristic of this class. From the Annals of the Harvard College Observatory, vol. 23, pt. II, 1901.

reversed E and F as better fitting a sequence. Class C was by then clearly seen to be the result of faulty plates. E and H were subsequently absorbed by other classes.

Once again, as the instrumentation and the quality of the spectrograms improved, a finer structure was necessary. The second and third of the trio, Antonia Maury and Annie Cannon, solved the problem in two quite different ways. In a study of northern stars in pt. 1 of vol. 28, 1897, Miss Maury developed an entirely new scheme, which placed the stars in 22 groups, identified by Roman numerals in parallel with the lettered sequence. Although the system had only a brief life, it had two important facets. First, her ordering showed that class B should precede A, and not be intermediate between A and F. She also suggested that her last group (XXII), related to bright-line class O, might come first in the sequence and be the connecting link to the emission-line nebulae. Second, Miss Maury appended additional descriptions pertaining to the appearance of the lines as well as to their total strength: 'a' stood for average spectra with well defined lines, and 'b' for those in which they were hazy and indistinct. Subclass 'c' was reserved for stars that exhibited narrow, sharp lines, or line-strength anomalies. The 'c' group was found later to consist of supergiants (Section 1.17), and the 'b' group to include stars in rapid rotation. The system was the first real attempt at 'two-dimensional' classification, categorization on the basis of two separate properties. Its full value was realized in 1907 when Ejnar Hertzsprung found that the 'c' stars had much smaller proper motions (Section 1.9) than did those in the 'a' and 'b' sets. This division was the initial evidence leading toward the discovery of the distinction between giants and dwarfs (Section 1.17), since from their slow apparent movements, the 'c' stars must be more distant and consequently more luminous than the average.

However, in a survey of southern stars published as part 2 of vol. 28, dated 1901, Miss Cannon added detail by simply decimalizing most of the Pickering series of letters. A star midway between K and M would be called K5M; the second letter was later dropped. She divided the O stars into groups Oa through Oe based on the character of their bright lines. Subtypes 'a' through 'c', in no way related to the Maury groups, denoted those with unusually broad and powerful such features, with 'd' and 'e' representing a more usual appearance with weaker emissions. Those with only dark lines she called Oe5. Letters a–c also served to classify the differences in the banded molecular spectra of M stars, with Md indicating the presence of emission lines. She also reversed A and B, as augered by Miss Maury's work, and from the presence of common lines, showed that O came before B. And thus out of these two studies was created the familiar and final form, OBAFGKM.

Miss Cannon's scheme was so elegant and versatile that it easily displaced Miss Maury's system. The sequence was not fully decimalized, nor is it still. But as the observations improve, it is simple to add more decimal compartments. For example, the B stars were easily subdivided into B0, 1, 2, 3, 5, 8, and 9; less structure was seen within G and K, which were assigned only G0, 5, and K0, 2, and 5. By the 1940s, B0.5, G2, G8, and K3 had been inserted. Uncertainty can be expressed simply by calling a star only A or F, without a number.

Although the order of letters stands alone without reference to physical explanation, the classifiers were well aware that it represented a temperature sequence.

THE HENRY DRAPER CATALOGUE.

39800   **5ʰ 49ᵐ.8**

H.D.	DM	R.A. 1900	Dec. 1900	Ptm	Ptg	Sp	Int	Rem	Pl. No.
		m.	° ′						
1	1055	49.8	+ 7 23	0.92	2.27	Ma	..	0,R	28,198
2	1211	49.8	+ 0 34	8.8	9.8	Ko	3	..	12754b
3	1114	49.8	− 0 14	9.1	9.1	A	3	..	12754b
4	2515	49.8	−36 17	8.5	9.4	Go	4	0,3	46181b
5	2273	49.8	−38 57	9.0	9.8	Go	1	..	46181b
6	2241	49.8	−39 32	8.4	9.5	K2	2	..	46181b
7	1925	49.8	−49 37	9.7	9.4	Fo	3	..	15220b
8	1968	49.8	−50 24	9.7	9.1	F5	2	0,2	24143b
9	534	49.8	−61 27	9.0	9.5	F5	4	..	15147b
10	418	49.8	−72 44	6.51	7.9	Ko	9	R	20540b
11	338	49.8	−73 30	10.2	10.6	F5	5	R	15167b
12	337	49.8	−73 31	10.2	10.6				
13	839	49.9	+61 7	8.7	9.3	Go	3	..	38154i
14	1333	49.9	+38 34	7.8	7.8	Ao	3	..	37429i
15	1193	49.9	+33 12	8.0	8.5	F8	3	..	37377i
16	..	49.9	+20 10	var.	var.	Md	..	R	286c
17	1212	49.9	−15 44	7.6	7.6	B9	9	..	20485b
18	3526	49.9	−24 20	10.1	10.4	Ko	1	..	45993b
19	2588	49.9	−29 2	9.8	9.1	Ko	2	..	42904b
20	2811	49.9	−31 43	8.4	9.9	K5	1	..	44364b
21	2464	49.9	−37 46	9.6	10.5	Ko	2	5,1	46181b
22	2070	49.9	−47 54	9.3	8.1	Ao	4	..	12756b
23	1926	49.9	−49 7	7.9	8.6	G5	4	..	12756b
24	1426	50.0	+49 59	9.4	9.8	F5	1	..	37366i
25	1324	50.0	+48 24	8.7	8.8	A2	2	..	37366i
26	1304	50.0	+36 8	8.6	8.7	A3	1	..	38124i
27	1079	50.0	+14 10	8.1	8.9	G5	2	..	37568i
28	950	50.0	+12 59	7.7	8.5	G5	4	..	37568i
29	970	50.0	+11 48	8.3	9.1	G5	2	..	38223i
30	1113	50.0	+ 8 40	9.3	9.3	Ao	2	..	38223i
31	1067	50.0	+ 6 43	8.8	8.9	A2	2	..	14071i
32	1074	50.0	+ 4 41	8.5	9.5	Ko	2	5,2	39866b
33	1115	50.0	− 0 30	7.8	8.4	Go	7	..	12754b
34	1116	50.0	− 0 57	8.3	9.1	G5	6	..	12754b
35	1210	50.0	− 7 24	9.1	9.5	F5	3	..	20546b
36	1211	50.0	− 7 32	9.1	9.1	Ao	4	..	20546b
37	1320	50.0	−11 50	8.6	8.7	A2	4	3,6	18414b
38	2697	50.0	−30 42	9.0	9.0	A2	5	..	44364b
39	2574	50.0	−35 50	9.4	10.5	Ko	1	..	46181b
40	2520	50.0	−36 52	10.7	9.9	F5	2	3,2	46181b
41	2206	50.0	−45 20	8.0	8.1	A2	7	..	12756b
42	491	50.0	−63 37	8.7	9.1	F5	4	..	15147b
43	461	50.0	−66 30	9.6	9.7	A2	2	..	18485b
44	463	50.0	−66 56	5.15	5.03	B5	..	0,7 R	28,198
45	1335	50.1	+38 16	7.24	8.42	K5	2	..	37429i
46	1287	50.1	+35 43	8.7	9.3	Go	2	..	38124i
47	1034	50.1	+25 19	7.71	7.77	A2	4	0,3 R	38084i
48	1117	50.1	− 0 15	9.6	9.6	A	2	..	12754b
49	1060	50.1	− 1 57	7.32	7.38	A2	7	0,4	12754b
50	1403	50.1	− 2 55	9.8	9.8	Ao	2	..	12754b

H.D.	DM	R.A. 1900	Dec. 1900	Ptm	Ptg	Sp	Int	Rem	Pl. No.
		m.	° ′						
51	1212	50.1	− 7 27	9.6	10.2	Go	1	..	20546b
52	1262	50.1	− 9 12	8.0	8.0	Ao	5	0,10	37625i
53	1321	50.1	−11 48	5.81	6.88	K2	7	..	20485b
54	1284	50.1	−17 47	8.2	9.2	Ko	4	..	12632b
55	1297	50.1	−19 43	7.34	7.8	G5	7	..	17395b
56	2764	50.1	−25 4	8.85	9.2	G5	2	..	12664b
57	2766	50.1	−25 56	9.6	9.6	Ao	2	..	12664b
58	2111	50.1	−43 18	11.0	11.2	G5	1	..	20649b
59	419	50.1	−72 1	10.1	11.3	K5	1	..	15167b
60	196	50.1	−79 34	9.3	9.7	F5	4	..	20557b
61	201	50.2	+81 31	8.9	9.5	Go	2	..	38330i
62	1133	50.2	+51 7	8.9	9.4	F8	2	..	37366i
63	1205	50.2	+45 53	6.56	7.34	G5	5	..	37391i
64	1288	50.2	+35 34	7.50	8.85	Mb	3	..	37377i
65	1045	50.2	+29 44	8.6	8.6	B8	3	..	37377i
66	952	50.2	+28 56	6.42	6.48	A2	6	R	37377i
67	1136	50.2	+19 21	8.1	8.1	B9	6	..	37568i
68	1213	50.2	+ 0 51	9.3	9.3	B9	3	..	12754b
69	1405	50.2	− 2 24	9.4	9.4	Ao	2	..	12754b
70	1213	50.2	− 7 1	9.6	10.0	F5	2	..	20546b
71	2768	50.2	−25 33	7.9	9.2	A3	3	..	12664b
72	2590	50.2	−29 47	9.1	10.1	K5	1	..	44364b
73	2698	50.2	−30 2	8.6	8.9	F5	5	..	44364b
74	2119	50.2	−41 59	8.0	8.3	Ao	7	..	20649b
75	799	50.2	−52 5	7.1	8.2	Ma	4	..	24143b
76	581	50.2	−58 45	9.4	10.0	G	2	..	18484b
77	357	50.2	−74 48	9.9	10.0	A2	4	..	15162b
78	1018	50.3	+52 2	9.2	9.5	Fo	2	..	37366i
79	1057	50.3	+17 59	8.3	9.3	Ko	1	..	37568i
80	1035	50.3	+13 59	8.5	9.3	G5	2	..	37568i
81	1036	50.3	+13 56	6.48	7.26	G5	..	5,7	56,81
82	951	50.3	+12 57	8.3	8.2	B5	5	..	37568i
83	1115	50.3	+ 8 58	8.5	9.1	Go	3	..	38412b
84	1075	50.3	+ 4 44	8.2	8.2	Ao	3	..	39866b
85	1286	50.3	− 4 4	8.5	9.5	Ko	4	..	20546b
86	1214	50.3	− 7 9	10.3	10.9	G	2	..	20546b
87	1215	50.3	− 7 10	10.3	10.9	G	1	R	20546b
88	1264	50.3	− 9 49	7.46	7.80	F2	9	..	20546b
89	1323	50.3	−11 20	9.1	9.2	A2	2	..	20581b
90	2592	50.3	−29 8	9.1	9.2	F5	3	..	42904b
91	2595	50.3	−29 10	6.17	7.3	F2	10	..	42904b
92	2113	50.3	−43 10	10.6	11.3	Ko	1	..	20649b
93	2026	50.3	−48 24	9.0	9.5	A2	5	..	12756b
94	419	50.4	+67 0	6.87	6.87	Ao	8	3,7	37545i
95	1308	50.4	+36 23	8.4	8.4	Ao	3	..	38124i
96	1178	50.4	+20 35	8.8	8.8	Ao	1	..	38084i
97	1022	50.4	+18 4	8.2	9.2	Ko	2	..	37568i
98	1061	50.4	+17 47	8.8	8.9	A2	3	..	37568i
99	1310	50.4	−12 57	7.78	8.28	F8	3	..	20485b
100	1228	50.4	−18 52	8.6	8.9	Fo	4	..	17395b

Figure 3.3. A page from the Henry Draper Catalogue. The bright star Betelgeuse (α Orionis, HD 39801) is the first entry at the upper left. The columns give HD: the Henry Draper number; DM: the Bonner Durchmusterung, or BD number (Section 1.2); R.A. and Dec: the coordinates, right ascension, and declination (Section 1.3); Ptm and Ptg: the visual and photographic magnitudes (Section 1.11); Sp: the spectral class; Int, Rem., Pl. No.: information on plate quality, references to remarks elsewhere in the volume, and the plate number. Note how much fainter red Betelgeuse (shown in Figure 1.1) is photographically than it is visually. Almost all of the stars in the table are well below naked eye visibility. From the Annals of the Harvard College Observatory, vol. 92, 1918.

Table 3.3. *The Harvard classes and their later development*

Type	Characteristic			Main sequence temperatures
O	He II, emission common			28 000–50 000 K
B	He I			9 900–28 000 K
A	H			7 400– 9 900 K
F	metals, H			6 000– 7 400 K
G	Ca II, metals			4 900– 6 000 K
K	Ca II, Ca I, molecules	R (CN, C_2)		3 500– 4 900 K
M	(TiO) S (ZrO)	N (C_2)	C	2 000– 3 500 K

Increasing carbon →

Development of class O

$\left.\begin{array}{l}\text{Oa}\\\text{Ob}\\\text{Oc}\end{array}\right\} \to \text{w} \to \left\{\begin{array}{l}\text{WC5–WC8}\quad\text{(carbon sequence)}\\\\\text{WN6–WN8}\quad\text{(nitrogen sequence)}\end{array}\right.$

$\left.\begin{array}{l}\text{Od}\\\text{Oe}\end{array}\right\} \to \left\{\begin{array}{l}\text{Oe}\quad\text{(H emission)}\\\text{Of}\quad\text{(He and N III emission)}\end{array}\right.$

Oe5 → O5–O9

Development of class M

Ma → M0–M2
Mb → M3–M5
Mc → M6–M8
Md → M0e–M8e

Premature theories of stellar evolution gave rise to the practice of calling the O stars 'early', 'and M stars 'late', 'a terminology still used: K0 is 'earlier' than K5. The ends of the sequence are open and can be expanded. In the years since the system was developed they have been extended to run from as early (hot) as O3 to as late (cool) as M10.

The work culminated in the magnificent *Henry Draper Catalogue* (Figure 3.3), published as *Annals* volumes 91 through 99 between 1918 and 1924 by Cannon and Pickering, in which Miss Cannon's classification of 225 300 stars was presented. Later work in the *Henry Draper Extension* brought the total to 359 082 stars by 1948. Lettered subdivisions of N, plus a new class R, which represents the carbon-branch version of the late G and K stars in the same way that N does for M (Figure 3.4), were added prior to publication. Type S, whose stars exhibit the oxide of zirconium rather than that of titanium, and which are really intermediate in carbon content between M and N, was added in volume 98. All the letters, identified by chief characteristic and temperature in Table 3.3, were now in place.

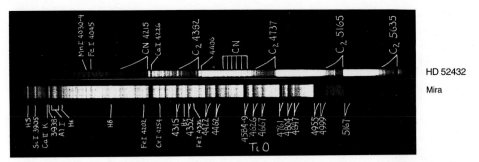

Figure 3.4. The dramatic differences among the cool stars are typified by the spectra of carbon stars (top, HD 52432, Secchi type IV, spectral class R5) and normal oxygen-rich M stars (lower, Mira, *o* Ceti, Secchi type III, class M7e). From *An Atlas of Stellar Spectra*, W. W. Morgan, P. C. Keenan, and E. Kellman, University of Chicago Press, Chicago, 1943.

3.4 Development and embellishment of the system

In light of new data, no system can be expected to remain unchanged. The original Harvard sequence continued to evolve to meet the needs of the times. Astronomers recognized very early the inconsistency in the treatment of the O, M, and N stars. In the 1920s, H. H. Plaskett began the reorganization of class O. Oa, b, and c, those with powerful emissions produced by circumstellar matter, were moved to the special category of *Wolf–Rayet stars*, which were eventually assigned decimal numbers in parallel nitrogen-rich and carbon-rich sequences, WN5 through 8 and WC6–8. The Oe5 group was seen to contain the assembly of normal O stars, and these were decimalized O5 to O9. The Od and Oe groups were considered normal, but with emission present, and were likewise rearranged as Oe and Of, where the e and f suffixes now respectively simply meant the presence of hydrogen or of helium and nitrogen emission lines.

Even Miss Cannon saw that the Ma, b, c stars formed a proper sequence, and these were rearranged into a numerical scale from M0 through M8. The Md stars, those with hydrogen emission, were arranged over the same numerical scale, but with *e* now appended to be consistent with the practice for the O stars. The M and O star development is outlined schematically in Table 3.3.

Figure 3.5 now displays the centerpiece of this chapter, the well known series of spectra taken in the 1930s by W. C. Rufus and R. H. Curtiss. The two panels show a sampling of the complete spectral sequence, with the chief features identified and the modern classes placed alongside. The fourteen spectra in Figure 3.5(*a*) show the progression of lines from B0 (ε Ori) to M7 (Mira). It is interesting to compare these with the older spectra of the same stars in Figure 3.2 taken with different equipment; the reader might try to identify the lines in the latter. Figure 3.5(*b*) shows the progression of, and the differences among, the O-type spectra, including the Wolf–Rayet stars. Their emission lines indicate strong outflows of gas; these rapidly evolving bodies will be examined in much more detail in Chapter 10.

a

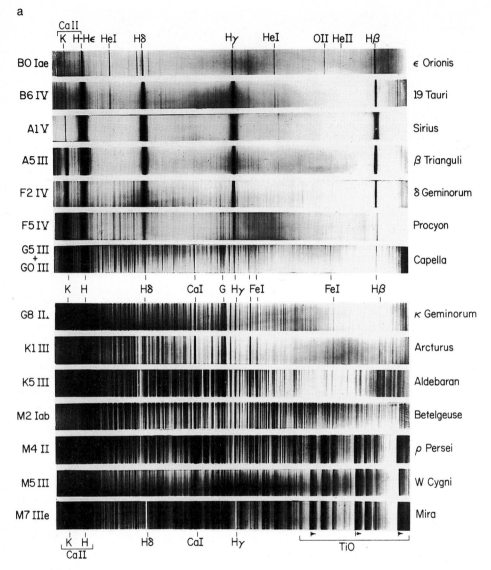

Figure 3.5(a).

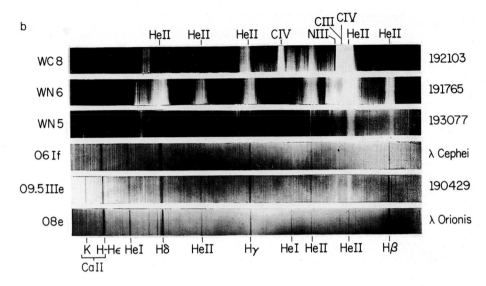

Figure 3.5. The spectral sequence. (*a*) shows representative stellar spectra in the blue and violet spectral region, obtained at the University of Michigan by W. C. Rufus and R. H. Curtiss. The Balmer series of hydrogen is conspicuous in the upper stars. The H and K lines at 3968 Å and 3934 Å are produced by ionized calcium and the G band arises from the CH molecule. Other identifications are He I at 4026 Å and 4472 Å; O II at 4649 Å; Ca I at 4227 Å; and Fe I at 4384 Å, 4405 Å, and 4668 Å. In the cool M stars the violet edges of molecular TiO bands lie at 4585 Å, 4762 Å, and 4954 Å. Classification is traditionally done in this blue-violet part of the spectrum. But the yellow-red can also be employed, and recently the far (satellite) ultraviolet and the infrared have become useful as well. (*b*) shows the spectra of the hot O stars, plus examples of carbon and nitrogen rich (WC and WN) Wolf–Rayet stars, which are often included in class O. Their emission lines show that they are surrounded by outflowing matter. The Observatory of the University of Michigan.

Other developments involved the considerable variances found among stellar spectra, and deviations from the so-called norm. In 1928, C. D. Shane reorganized the classification of the carbon stars (Figure 3.4 and Table 3.4). He renumbered the R subdivisions continuously from 0 to 9, and changed the Na, b, c to N0 through N7, giving detailed criteria. The final touch was added in 1941 when P. C. Keenan and W. W. Morgan re-ordered all the carbon stars, both R and N, into one grand sequence C0 through C9, which crudely parallels the normal or oxygen-rich variety from G4 to late M (see Table 3.4).

As the science of spectroscopy improved, astronomers also became much more aware that stars of a given class could have some very different characteristics: broad lines or narrow, as found by Miss Maury, odd line strengths for various elements, emission lines, and more. The greatest distinction is probably the separation of the stars into giants and dwarfs, as indicated first by Hertzsprung, and lucidly demonstrated by Henry Norris Russell in 1914 in what was later to be called the Hertzsprung–Russell or HR diagram (Figure 3.6). In this array, absolute magnitude or luminosity is plotted

Table 3.4. *Development of carbon star classes*

Harvard	Shane	KM	Equivalent normal star[a]
	R0	C0	G4–G6
R0	R1	C1	G7–G8
	R2	C2	G9–K0
R3	R3	C3	K1–K2
R5	R4		
	R5	C4	K3–K4
	R6		
	R7		
R8	R8	C5	K5–M0
	R9		
	N0	C6	M1–M2
Na	N1		
	N2	C7	M3–M4
Nb	N3		
	N4	C8	
	N5		
Nc	N6		
	N7	C9	

[a] Original equivalents published by Keenan and Morgan. They are only approximate, and for the later classes very much so. There is good evidence that the N stars correspond to M classes later than given here. The C-type responds to both carbon abundance and temperature and consequently there is a large and essentially unknown temperature range for each carbon subclass.

against spectral class, and the stars are clearly seen to fall into broad bands that represent the dwarfs, or *main sequence*, and the more luminous *giants* and *supergiants* (Section 1.17). A giant can be hundreds of times larger and brighter than its dwarf counterpart, and the vastly lowered atmospheric density causes a very noticeable change in the ionization balance and thus in the overall appearance of the spectrum. In the earlier-type dwarfs and even in the giants, the densities are sufficiently high to cause an appreciable broadening of the hydrogen lines through the effects of atomic collisions on the energy levels (Section 2.8). But in the supergiants, a more extreme breed, the densities are so low that the lines become very narrow, and we see Miss Maury's c-stars.

In the 1920s, in response to the better data, a system of prefixes and suffixes to the Harvard classes was generally adopted to indicate the numerous observed variations.

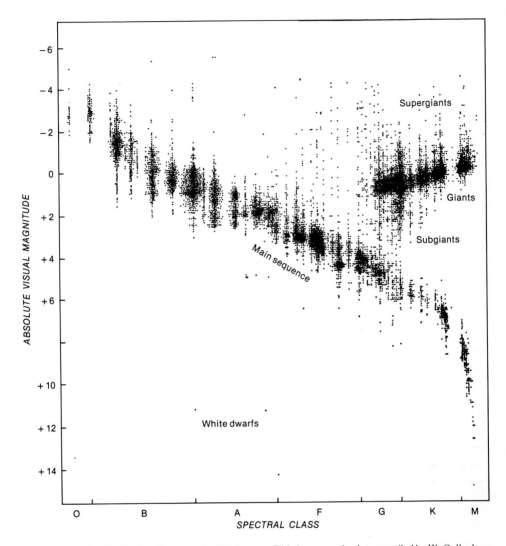

Figure 3.6. The distribution of stars on the HR diagram. This famous early plot, compiled by W. Gyllenberg of the Lund (Sweden) Observatory, beautifully illustrates the separation of the giants from the dwarfs. A number of supergiants scatter across the top. Other zones (subgiants and white dwarfs) are discussed later in this chapter. The banded structure is caused by the stars being placed in the discrete bins of their assigned spectral types. The lower main sequence (K and M) is much more heavily populated than shown here; at the time the drawing was made these dim stars, and the white dwarfs, were difficult to observe and were very much under-represented. The sparsely populated separation between the giants and dwarfs is the renowned 'Hertzsprung gap', which is now recognized to extend and expand in size upwards from the main sequence, roughly centered in the neighborhood of class G. It is a product of the timescales of evolution. Diagram adapted from *Stellar Evolution: An Explanation from the Observatory*, by O. Struve, Princeton University Press, Princeton, 1950, 1978. Reprinted with permission of Princeton University Press.

Table 3.5. *Prefixes and suffixes to the spectral classes*

The terms below give descriptions that are used with the spectral classes to provide fine structure. Obsolete terms are in italics.

Prefixes

b (*ab*)	*wide lines*	d	dwarf (main sequence)
a (*ac*)	*normal lines*	g	giant
c	sharp lines (supergiant)	sd	subdwarf
		wd =D	white dwarf

Suffixes

e	emission (H emission for O stars)
em	emission by metal lines
er	reversed emission (central absorption within the emission line)
ep	peculiar emission
eq	P Cyg emission (emission with an absorption component toward shorter wavelength)
f	He and N emission for O stars
n	diffuse lines
nn	very diffuse lines
s	sharp lines (other than Maury's 'c')
k	interstellar lines
v	variation in the spectrum other than that caused by velocity effects
p = pec	peculiar spectrum
m	strong metallic absorption
wk	weak lines
!	marked characteristics

Some, like Miss Maury's 'b' and 'a', are defunct. Others, like 'e' and 'f' for emission, are in active use today. Several more, while not as common, still appear regularly in the literature, and all astronomers should be aware of their meanings. They are described in detail in Table 3.5. Their use is simple: a giant M7 star with emission would be called gM7e, and an A0 supergiant with some highly unusual line characteristics, spectral variation and interstellar lines, might be cA0pvk. Modifiers such as 'p' (for peculiar) follow the letter that states a given property.

3.5 Two-dimensional classification

By the late 1930s it was again time for major reorganization of the standard sequence. The system of adjunct lower-case letters was insufficiently fine to describe a

Table 3.6. *The MK luminosity classes*

The table gives a brief description of the luminosity classes and their absolute visual magnitudes at various spectral types. The letters *a* and *b* are sometimes suffixed to the classes to indicate whether a star is on the bright or dim side of the average, as for class I. Anomalously high abundances are indicated by appending the relevant chemical or molecular symbol; sometimes a '+' or '−' and a numerical index is added to the symbol to show enrichment or depletion: see Section 3.6.

		Absolute visual mag		
	Spectral type	B0	F0	M0
0 (zero)	The extreme, luminous supergiants		−9	
	The Magellanic Clouds and the Galaxy			
Ia	Luminous supergiants	−6.7	−8.2	−7.5
Ib	Less luminous supergiants	−6.1	−4.7	−4.6
II	Bright giants	−5.4	−2.3	−2.3
III	Normal giants	−5.0	1.2	−0.4
IV	Subgiants	−4.7	2.0	
V	Main sequence	−4.1	2.6	9.0
sd (VI*)	Subdwarfs			10
D, wd (VII*)	White dwarfs	10.2	12.9	

* Infrequent

continuum of spectral characteristics related to luminosity in the way that spectral types referred to temperature. The work was completed in 1943 by Morgan, Keenan, and E. Kellman, who defined a sequence of six luminosity classes (Table 3.6), and re-defined each Harvard subclass by the spectrum of a selected star in close parallel with the Draper descriptions. Luminosity classes Ia and Ib are the two supergiant divisions, III represents the giants, and V the main sequence (the dwarfs of the older parlance). Number II, usually called bright giants, describes stars intermediate between Ib and III. Class IV stars fall between the giants and main sequence, and are termed subgiants. Finer structure is rendered by using composite classes (for example III–IV), although sometimes other methods such as the 'a' and 'b' suffixes of luminosity class I are employed (i.e. stars of type IIIa lie above the average of the giants on the HR diagram, with those of IIIb below). The luminosity class is suffixed to the spectral class, so that a bright F0 supergiant would be called F0 Ia. The other descriptive suffixes of Table 3.5 then usually follow the luminosity class. A B3 giant with emission lines would then be called B3 IIIe, and a bright O5 supergiant with helium and nitrogen emission would be labeled O5 Iaf. These luminosity indicators are given for all the stars of Figure 3.5. Notice that almost all are giants and supergiants, reflecting the predominance of these brilliant stars in the naked-eye sky.

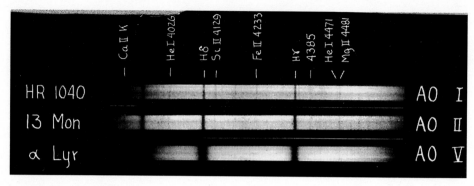

Figure 3.7. An example of luminosity classification. The hydrogen lines are much narrower in supergiants than in main sequence stars (dwarfs) because of lower atmospheric pressures. Other criteria are used for different Draper classes. From *An Atlas of Stellar Spectra*, by W. W. Morgan, P. C. Keenan, and E. Kellman, University of Chicago Press, Chicago, 1943.

In this MKK (later, MK) system, the authors relied heavily upon extensive previous work done at the Mt. Wilson Observatory and elsewhere, in which the astronomers demonstrated how various lines and line ratios change with absolute magnitude. For example, with increasing luminosity, the strength of the cyanogen (CN) bands greatly increase at late G and early K, that of the neutral calcium Fraunhofer 'g' line weakens markedly at early M, and the hydrogen lines narrow at type A (Figure 3.7). The luminosity class is assigned by taking account of a number of such criteria, which we will inspect in later chapters. The loci of these six luminosity classes are shown in the skeletal HR diagram in Figure 3.8. Notice that the absolute magnitude of a class depends upon the spectral type, and that all converge toward the upper left.

Three other luminosity classes and loci remain to be considered. Many years ago, astronomers found that the brightest supergiants in the Magellanic Clouds outshone those in our Galaxy. These are occasionally called class 0 (zero, not be confused with Draper O stars), a term sometimes applied to the most luminous galactic stars as well. They are also infrequently referred to as super-supergiants, hypergiants, or as 'S-Doradus type'. The lesser members of the class are typed as Ia-0.

Then there is an interesting group of stars called *subdwarfs*, prefixed 'sd'. These lie to the left of the main sequence roughly from F through K, and are the result of a comparatively low metal content in their atmospheres. The weaker lines that ensue make them appear to be somewhat earlier than expected for a normal dwarf at a

Figure 3.8. The Hertzsprung–Russell (HR) diagram, showing the average locations of stars of the various luminosity classes. The left-hand axis is absolute visual magnitude, M_v (Section 1.13). The Sun lies at $M_v = 4.8$. The stars actually occupy broad bands centered on the lines, as seen in Figure 3.6. The giant and supergiant zones are especially broad. Diagram by the author, compiled from data presented by: R. H. Allen in *Astrophysical Quantities*, 3rd edn., Athlone Press, London 1973; T. Schmidt-Kaler in *Landolt–Börnstein Tables*, Group VI, Vol. I, Springer, New York; and R. M. Humphries and D. B. McElroy, in an article in the *Astrophysical Journal*.

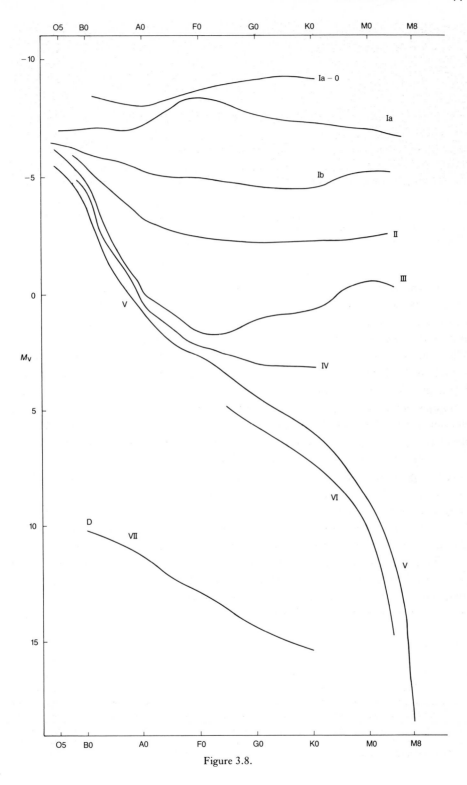

Figure 3.8.

particular absolute magnitude. The subdwarfs are sometimes called 'class VI', and are known to belong to the metal deficient Population II halo of our Galaxy (Section 1.8).

Finally, we encounter a dim set of stars, the white dwarfs (Section 1.17), some 10 or more magnitudes below the main sequence. Although they range through most of the spectral sequence, the first ones found (the companion to Sirius for example) were white, and the name is now applied generally to all of them. These carry prefixes of 'wd' or 'D', and are the end products of normal stellar evolution. Other zones and branches exist as well, which we will examine in appropriate chapters below.

The immense task of reclassifying the HD catalogue stars on the MK system has been undertaken at the University of Michigan by N. Houk. Their Curtis–Schmidt telescope was moved to Cerro Tololo in Chile, and the astronomers are systematically observing toward northerly declinations. Additional plates are being taken with the Warner and Swasey Observatory's Burrell Schmidt at Kitt Peak. This work will undoubtedly be for the next century what the Draper Memorial has been to ours.

3.6 The third dimension

The subdwarfs, with their deviant chemical compositions, give us an indication of the direction in which we must progress in the classification of stars. The spectral sequence, with its luminosity classes, is primarily defined only for the Population I stars that inhabit the galactic disk, and which have chemical compositions similar to that of the Sun. There are, however, some fascinating variations from this norm, which can make classification based upon population I criteria quite uncertain. We must thus expand the system to encompass the dimension of chemical abundance differences, about which we yet have much to learn.

Deviation from the solar mixture of elements is the result of the state of the Galaxy at the time of formation of the star, nuclear processes that change the composition over the star's lifetime, mixing of surface gases with those of the nuclear burning core, the influence of a close binary companion, and other causes. Variations manifest themselves differently within each spectral class. At M, for example, we have the carbon-rich N and C stars, and the intermediate-carbon S class that we discussed earlier. At class K we see barium-rich stars; at G and K we see variations in CH and CN molecular line strengths; at A, exotic metals may make an appearance; and of course on the lower main sequence we have the metal-deficient subdwarfs.

The parallelism among these various kinds of stars is still highly uncertain. We will look more closely at these numerous divergences from the ordinary in the chapters that follow. Classification is sometimes indicated by the symbol of the deviant element or molecule together with a strength index appended to the MK type. Examples are 40 Orionis, class K0 III CN-2, which shows that the CN molecular feature is weaker than expected for this class, and 36 Aurigae, B9.5p Si Fe, which has a peculiar spectrum and overly strong silicon and iron lines. But as yet we have not found a logical and simple way of incorporating all these changes into a standard system. Clearly, a great amount of work remains to be done in a field that has now been active for over 150 years.

3.7 Physical basis of the sequence

The HR diagram is simply sketched out in Figure 3.8 to show the locations of stars with respect to spectral type and luminosity class. So far, we have needed to make little reference to physical quantities and how they might vary across the surface of this graph, which is, of course, the basic idea of empirical classification. That subject lies within the realm of astrophysics, and is one with which we have been coping for much of this century.

We have learned that the spectral sequence is a result almost solely of a temperature progression among stars, which runs from a high of about 50 000 K at very early O to a low near 2000 K at late M. Temperature controls ionization and electronic excitation among the energy levels (Sections 2.6, 2.7, and 2.11); the faster the atoms are moving in the stellar atmosphere, the more electrons that will be knocked away, that is, provided with energies above the ionization limits. At the low temperatures of the M stars most of the metals will be in their neutral states: we will see Fe I, and the Ca I g-line will be very strong. Even though neutral hydrogen dominates everything, it will be nearly undetectable because so few electrons can be pumped to the second level (see Section 2.11). In addition, molecules such as TiO (and ZrO in S stars, or CO in type N) will be abundant.

As the temperature climbs, the metal atoms become ionized: different ones at different spectral types, depending upon their individual ionization potentials (the minimum energies required to knock the electrons away). Ca II, the spectrum of Ca^+, strengthens in classes K and G at the expense of Ca I; Fe I converts to Fe II in class G, then into Fe III in F. Hard-to-ionize atoms like silicon do not enter their first ionization stage until we reach stars as hot as class A, and helium not until class B.

Within a given ionization stage the line strengths still vary with temperature, as the population of any particular upper energy level (the ground state is more or less immune) depends critically upon the energy of the gas and its ability to excite the electrons. Hydrogen, discussed in detail in Section 2.11, provides the most obvious example. This atom does not ionize at all readily, so that the Balmer lines simply and steadily increase in strength through most of the sequence as level 2 becomes more populated.

But near A0 ionization sets in and these lines then rapidly weaken. Class A leads the original Draper list because of the powerful hydrogen lines; thus when we re-order by temperature (from high to low), B and O, with progressively weaker Balmer absorptions, must physically come before A: a fact that can be inferred from spectral continuity alone without reference to temperature. The sequencing itself led the way to our physical understanding. These ionic progressions are schematically graphed in Figure 3.9.

A complication arises when we compare luminosity classes with one another. For a given spectral type, (or broadly speaking, ionization balance), the temperatures of giants and supergiants are generally, especially among later types, progressively lower than are those of their main sequence counterparts. The luminous stars, as pointed out before, have lower atmospheric densities because of their larger sizes. The level of

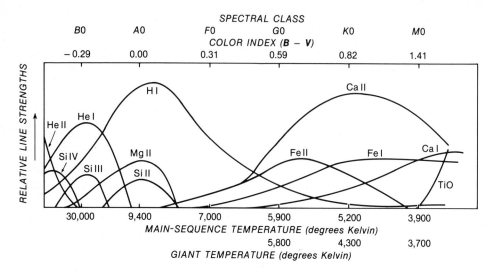

Figure 3.9. The relative variation of line strengths with spectral class. Hydrogen is strongest at type A, neutral calcium and TiO at late M. Values of color index (B–V, Section 1.14) for main sequence stars and temperatures for both these and giants are also given. Figure 3.10 provides a finer conversion among the different variables. From an article by T. Schmidt-Kaler in the *Landolt–Börnstein Tables*, Group VI, Vol. I, Springer, New York.

ionization in a gas is primarily dependent upon temperature, but is also responsive to density or pressure. The numerical ratio of the number of a certain ion to the one below it (Fe^{+2} to Fe^+ for example) depends upon the competition between the processes that cause ionization (collisions of all kinds and radiation) and those that promote recombination (electron–ion collisions). As we reduce the pressure and increase the spacing between atoms, the recombination rate is reduced relative to the ionization rate and the level of ionization goes up. To maintain the original ionization level the temperature must then be reduced as well. The differences are not large, only a few hundred degrees, but they are important for theoretical analyses. The matter is really more complex than just described however since at classes A and F the more luminous stars become hotter than their dwarf analogues, in part because of the specific spectral criteria used to define the different classes.

We now see that there are three ways of treating ionization in a star: the spectral class, the temperature, and color index (Section 1.14), principally B–V. Any of the three may be used, depending on the problem at hand. The conversions and calibrations, the results of decades of observation, are shown in the graph of Figure 3.10. Remember, however, that we defined two measures of color, B–V and U–B. For completeness sake the relation between the two is shown in Figure 3.11; it will become important at a later time.

The classical HR diagram, which plots absolute magnitude against spectral class, can then be recast as a 'color–magnitude diagram', or as a distribution of stars on the 'log

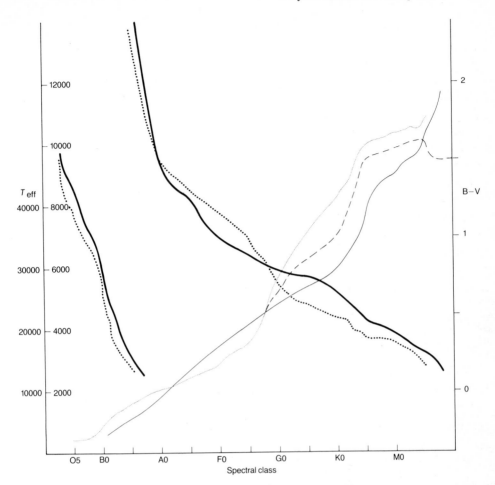

Figure 3.10. The relation between spectral class, effective temperature (T_{eff}) and the B–V color index. The main sequence, giants, and supergiants are respectively indicated by solid, dashed, and dotted lines. The heavier lines refer to temperature; the left-hand curves go with the higher-temperature left-hand scale, and the right-hand curves go with the lower-temperature right-hand scale. The lighter color-plots go with the B–V scale on the far right. Effective temperatures are for supergiants in general (average of Ia and Ib), the colors are specifically for the Ia variety. Earlier than G0 the giant temperatures fall closer to those of the main sequence; later than that they are more like those of the supergiants. Giant colors earlier than F8 are similar to those of the dwarfs. Over most of the spectral sequence the higher luminosity stars have cooler temperatures and redder colors than the dwarfs because of their lowered atmospheric pressures. This relation reverses over a short spectral range between A0 and F8. The reversal of the color-spectral class correlation for late-type giants is caused by the development of powerful molecular bands. Diagram by the author. Temperature data from articles by R. M. Humphries and D. B. McElroy in the *Astrophysical Journal* and by D. M. Popper in *Annual Review of Astronomy and Astrophysics*; color data from T. Schmidt-Kaler in *Landolt–Börnstein Tables*, Group VI, Vol. IIb, Springer, New York 1982.

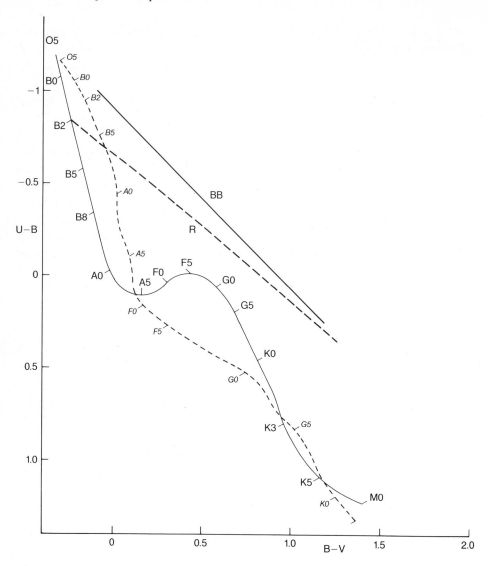

Figure 3.11. The color–color diagram, which shows U−B plotted against B−V for main sequence stars and for supergiants, with various spectral classes indicated; supergiant classes are in italics. The curve for a blackbody is indicated by a solid line labeled 'BB'. Clearly, the stars do not behave like blackbodies, the deviation being caused by atmospheric transparencies that change with wavelength. The hook in the main sequence curve between A5 and G0 is caused by the prominence of the Balmer continuum, which influences U. Its absence in the supergiants is related to the weakness of the hydrogen absorption in these luminous stars (see Figure 3.7). The dashed curve labeled R shows what happens to the colors of a B2 star that is subject to the dimming effects of interstellar dust: see Chapter 9. From data presented by T. Schmidt-Kaler, in *Landolt–Börnstein Tables*, Group VI, Vol. IIb, Springer, New York 1982.

L–log T plane' in which the logarithms of stellar luminosities are plotted against those of their temperatures. The resulting distributions will look somewhat different from the standard HR diagram because of the conversions of Figure 3.10, and in the differences in the temperature sequences among luminosity classes.

Finally, recall from Chapter 1 that the main sequence represents a continuous *mass* sequence. The hottest stars may have masses of 100 Suns or more, the coolest no more than a tenth, although the true upper limit set by star-formation processes is not really known. The distribution of the stars across the HR diagram into the other luminosity classes we have discovered to be the result of the processes of stellar evolution that were developed briefly in Section 1.17: the natural aging effects that take place when the nuclear fuels that power the interiors of stars run out. Many critically important steps in the process, though, are not understood, and we must rely on empirical classification to guide our studies. Now, with this background, we proceed with a closer examination of each spectral type, where we will look at the stars' physical natures, and fill in the HR diagram to show the details about the various kinds that occupy our Galaxy.

4 *The M stars: red supergiants to dwarfs*

As Orion sets, the scorpion rises. The gods, we read, placed the hunter in the sky such that he need not look upon his slayer, Scorpio. It is perhaps ironic then that these two constellations contain the only first magnitude stars of class M, the coolest of spectral types. We can almost always see one or the other of these bright red jewels, Betelgeuse (α Orionis) or Antares (α Scorpii) as they wheel around us on opposite sides of the sky, placed respectively within the winter and summer segments of the Milky Way. They are each shown within their constellations in Figures 1.1 and 4.1.

More than any other, type M epitomizes the vast differences found among stars: the brightest of them are *50 billion* times more luminous than the dimmest (Figure 4.2). *All* of the red stars we can see with the unaided eye are giants or supergiants; *no* dwarf M can be seen without a telescope. That is not to imply that the faint dwarfs are of lesser importance: both broad groups have their own special significance. The giants, and especially the supergiants, are enormously distended, with dimensions on a solar system scale; they are rare and ephemeral – they have recently evolved from much hotter main sequence stars, and will shortly, perhaps within ten to a hundred million years, change into something quite different; they lose their mass at high rates, and commonly vary in brightness over long periods of time. The dwarfs, in contrast, are amazingly small, more comparable with the planet Jupiter in radius, and their real claim to distinction is their number: there are more main sequence M stars than any other kind. Their feeble luminosities are due to very slow rates of energy production that give them enormously long lifetimes: an M dwarf born even at the time of the origin of our galaxy some 13 billion years ago will even now have suffered little change.

4.1 The spectra of the coolest stars

We will return to the above characteristics, and how we have learned of them, but first let us examine the font of almost all our knowledge, the spectrum. These coolest of stars have by far the most complex of all spectra, as the reader has already seen from the Figures 3.2 and 3.5. Those of Betelgeuse (Figure 4.3) and of the famous long-period-variable Mira (*o* Ceti, M7e, which changes from third to tenth magnitude with a period of a year or so: see Section 4.5 ahead), are chopped up by literally thousands of individual absorption lines, many very powerful and deep, others seen only at the highest dispersions. In some spectral regions, especially for coolest stars, there are so

Figure 4.1. Northern Scorpio. The supergiant Antares (M2 Ia) is just below the left-most of the three nebulae, just up and to the left of center. The head of the Scorpion, which consists of B stars, can be seen at the upper right. Antares appears relatively faint because it is red and does not record well on the blue sensitive photographic plate. From *An Atlas of the Milky Way*, by F. E. Ross and M. R. Calvert, University of Chicago Press, Chicago, 1934.

many lines that what little is left of the background continuum peeks through, seeming to masquerade as emission lines. It is a truly imposing task, still far from complete, to identify all these features with known elements and molecules.

It is the molecules (Section 2.10) that give the M spectra their very characteristic banded appearance. They are easily broken up by collisions with their neighbors in the gas, and consequently can form only at lower temperatures. Just a few hardy ones like CN, CH, and CO exist in our G2 sun; and as we progress to later types, more can form. At K5 the molecule that dominates the M-star spectra, titanium oxide, TiO, begins to be seen. More fragile species – VO (vanadium oxide), MgH, H_2 – develop toward lower temperatures as TiO strengthens, and by late M whole blocks of the spectrum may be removed by molecular absorption. Figure 4.4 displays the spectra of three M giants, showing the development of these features and the extraordinary absorbing power of

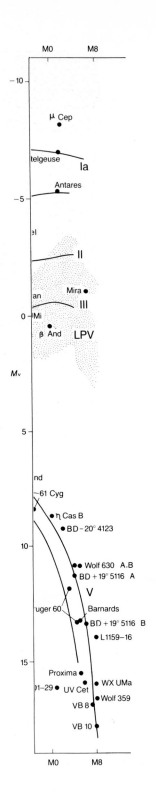

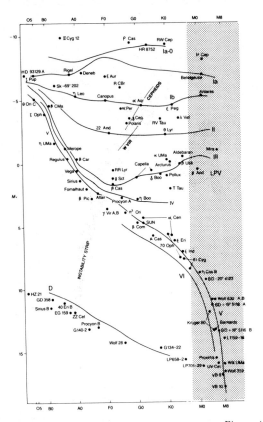

Figure 4.2.

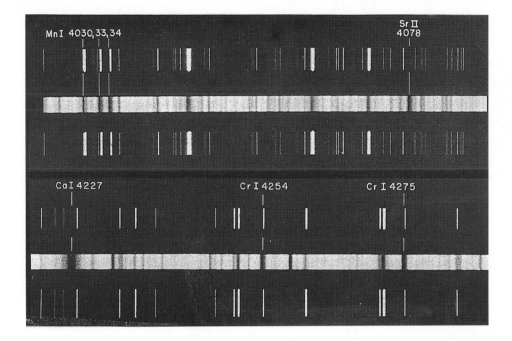

Figure 4.3. Two sections of a very high dispersion spectrogram (the light strips) of Betelgeuse (α Ori), which show a large number of neutral lines; the Ca I λ4227 feature is Fraunhofer 'g'. Most of the other strong absorptions are Fe I. Compare this spectrum of this star with the previous one in Figure 4.4 to see the great importance of dispersion in the exhibition of fine detail. The sets of bright-line features that flank each stellar spectrum are produced by the comparison arc in the spectrograph that provides reference positions and wavelengths used for measurement of the absorption lines (see Section 2.14). From an article by A. J. Deutsch in *Stellar Atmospheres*, J. L. Greenstein, ed., University of Chicago Press, Chicago, 1960.

TiO, which literally hides and destroys the background continuum. At M8 there is nearly nothing left to see in the blue-violet. These graphical spectra are not photometrically correct, that is they do not show the true distribution of energy with color; they are distorted because of large variations with wavelength in the sensitivity of the detector, reflectivity of the telescope optics, and atmospheric transparency. Nevertheless it is easy to see that, especially for the late types, the spectrum does not even come close to resembling that of a blackbody, grossly distorting the color indices. Superimposed upon this complex are thousands of atomic lines of the common metals, iron, chromium, manganese, and the like.

Figure 4.2. The HR diagram for class M, with several prominent stars that indicate the huge range of luminosity. The brightest illustrated here, μ Cephei, is nearly as large as Uranus' orbit, and is 50 billion times brighter than VB 10 at the bottom of the main sequence, which is only a few times larger than the Earth. The shaded area shows the location of the long period (Mira) variables. In succeeding chapters we will move this strip across the HR diagram, illustrating the class under discussion. See Figures 3.8 and 12.7 for credits.

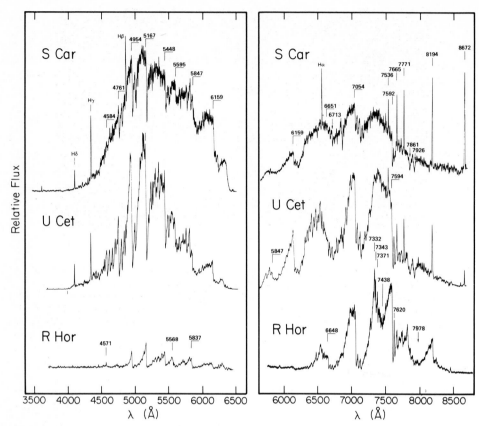

Figure 4.4. The spectra of three red giants from the near ultraviolet to the near infrared, showing the progression and deepening of molecular absorptions. All are long period Mira-type variables (Section 4.5). Top: S Carinae, M2e III (which varies from $m = 5$ to 10 over a 150 day period); middle: U Ceti, M4e III (7th to 13th magnitude, period 235 days); bottom: R Horologii, M8e III (5th to 14th magnitude, period 400 days). The spectral classes of Miras change during the variation cycle. Those given here are appropriate to the time at which the spectrogram was made. The heads of TiO bands are illustrated by wavelength and a right-angled bar; those without heads are noted by a downward arrow. The hydrogen emission lines (Hα, Hβ, Hγ, and Hδ) are typical of the long-period Mira variables (see Section 4.5) and indicate shock waves moving outward through the low density stellar atmosphere. In the coolest stars the temperature is so low, and the molecular bands so powerful, that there is little to see in the blue–violet spectrum. From an article in the *Astronomical Journal*, by L. Celis S.

The spectra are very easily classified M0–M8 on the strengths of the various bands (Figures 4.4 and 4.5). Other compounds that have features in the infrared part of the spectrum are similarly used. At the coolest end of the sequence there is little light emitted in the blue portion of the spectrum traditionally used for classification, and the subtype is better determined by using the strengths of the red TiO bands.

As we saw in the last chapter (Section 3.5) we can use various features in the spectra to separate the luminosity classes from one another. The specific lines that are employed change as we proceed along the spectral sequence. For example, in class M we can use the strength of the neutral calcium g-line at $\lambda4227$ Å, which increases in strength only slowly with temperature. It is, however, very sensitive to pressure, and

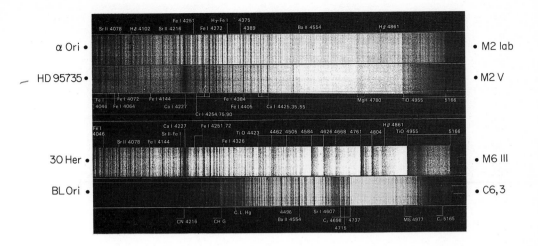

Figure 4.5. The spectra of three M stars and a carbon variable show dramatic differences. The M2 dwarf, HD 95735, has about the same TiO band strength as the M2 supergiant Betelgeuse (α Ori), but because of luminosity-dependent effects it has a much stronger Ca I feature and no visible hydrogen lines. The M6 giant 30 Herculis has far more powerful TiO bands than either, a result of its low temperature. The C6 giant BL Orionis is comparable in temperature to Betelgeuse, but shows strong carbon bands and no TiO. The number '3' following the C6 spectral designation is an index that illustrates carbon line strength. Note the several iron and strontium lines. From *An Atlas of Representative Stellar Spectra*, by Y. Yamashita, K. Nariai, and Y. Norimoto, University of Tokyo Press, Tokyo; John Wiley & Sons, New York, 1978.

exhibits what is called a *negative luminosity effect*, that is its strength varies inversely with stellar brightness, becoming very weak in the spectra of the brilliant low density giants and supergiants. The hydrogen lines, barely visible in early M giants, behave oppositely, disappearing altogether as the dimmer stars of the main sequence are approached. They are said to display a *positive* luminosity effect, their strengths changing in concert with brightness. Several iron line ratios are useful as well.

4.2 Carbon stars

Although this chapter is entitled 'The M Stars', we cannot consider them apart from the parallel sequences of types N, or C, and S, which are all giants, and which represent deviations in chemical composition. With maximum simplicity, we can divide the cool giants into two broad groups, oxygen- or carbon-rich (Figure 4.5), depending upon which element dominates the other, epitomized by Secchi's types III and IV. Carbon has a great affinity for oxygen. If oxygen is more abundant, some will be used in making CO and the like, which is seen in M stars, but most of it is available to produce metallic oxides. However, if carbon dominates, it will use up all the oxygen, mostly through CO production, and then develop a variety of other carbon compounds. Thus the TiO so characteristic of type M will be replaced by the C_2 that is notable in class C or N. Other simple compounds such as C_3, CH, and CN are seen as well. The resulting spectra are so much more complex than those of the M stars that accurate classification is often quite difficult, if not impossible, which accounts for the repeated reorganizations of stellar types described in the last chapter.

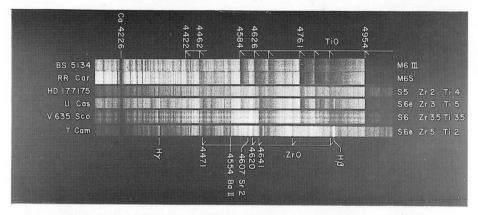

Figure 4.6. Six spectra that show the transition from pure type M with strong TiO to nearly pure S with strong ZrO. The numbers after the symbols for zirconium and titanium are indices that represent the relative line strengths. The star RR Car is intermediate between classes M and S. From *An Atlas of Spectra of the Cooler Stars*, by P. C. Keenan and R. C. McNeil, Ohio State University Press, 1976.

In reality, these stars do not divide simply into two discrete groups, but fall into an almost continuous sequence in carbon-to-oxygen ratio (C/O), which encompasses the S stars as well, in which ZrO is so prominent (Figure 4.6). Zirconium has a stronger affinity for oxygen than titanium, but is much less abundant, so in normal M stars the ZrO lines are weak or absent. But as C/O increases, and the C grabs the O, what little is left will go the zirconium, which itself is increasing in abundance along with the carbon. By the point at which carbon is about equal to oxygen in abundance, the spectrum of TiO has weakened considerably, and that of ZrO predominates. The presence of oxides of other heavy elements such as yttrium and vanadium (YO and VO) accompany the change. Intermediate steps, the MS (between classical M and S) and the SC stars can easily be recognized (Figure 4.6).

C/O variations among the C stars are often expressed through the classification, for example (C6, 1), or (C7, 9), where the first number is the temperature sub-class and the second refers to the overall strength of the carbon bands. Thus we see here at least a qualitative explanation of the branching of the spectral sequence, which extends to the difference between types K and R.

The carbon stars are all giants, so there is no need for luminosity criteria. The peculiar abundances of carbon and other elements arise from internal processes that bring by-products of nuclear fusion to the stellar surface. These stars represent the clearest example of the introduction of the third dimension to spectral typing, so necessary for the progress of the subject.

4.3 Temperature and color

The calibration of the spectral sequence in terms of effective temperature (that required for a perfect blackbody with a radius and total luminosity equivalent to the real star; Section 1.15) is a fundamental astrophysical problem. To find it properly, we need to measure a star's distance (to get luminosity, L) and radius, R. Then, from Section

1.15, T_{eff} is proportional to the fourth root of L divided by the square root of R ($T_{eff}^4 = L/4\pi\sigma R^2$). However, both these quantities are known from direct determinations for only a handful of stars, so we must seek other means.

In principal, a straightforward way of finding the effective temperature would be to fit the best possible blackbody curve (Figure 1.11) to the observed continuous spectrum. For hot stars with simple spectra this method actually works rather well, but in general it does not. The problem, as we have noted, is that the transparency of the stellar atmosphere changes with wavelength, so that we see to different depths and temperatures in different parts of the spectrum, that is, the observed spectrum is actually a superposition of many blackbody curves. In addition, at the cool end of the spectral sequence, there are so many absorptions that the continuum can hardly be *found*, let alone measured and fitted. Nevertheless, such measurements combined with proper mathematical averaging of the data can yield a reasonably good approximation of T_{eff}.

A measure of effective temperature can also be achieved through the application of atomic principles to the observed spectrum. We may, for example, compare with theoretical prediction the strengths of two or more absorption lines of the same ion that arise from different energy levels, and that have different sensitivities to temperature. At the extreme we can construct a mathematical model of the star to find how temperature changes with atmospheric depth. Different lines are formed in the different layers (see Section 2.11). We determine the temperatures by adjusting them until we can match not only the observed continuum, but so that we can also calculate accurately the strengths of all the absorption features that lie within large blocks of the spectrum. This information can in turn be converted into the effective temperature that we ought to find could we actually measure stellar luminosity and diameter. The procedure is obviously very time-consuming, and depends upon electron transition probabilities and chemical compositions that may not be well known. In fact this method is often used to derive temperature and the atomic abundances simultaneously, since the latter may be adjusted as well in order to achieve a satisfactory fit between theory and the data.

The results of such studies range from reasonably satisfactory for the earlier M stars to very poor for the later types. The temperatures of the M dwarfs range from 3900 K at M0 to 2600 K at M8. As pointed out in Chapter 2, for a given spectral class, giants are both cooler and redder. Low density enhances ionization and suppresses molecular formation, which for a given temperature results in an earlier spectral type. In order for a giant with its rarefied atmosphere to have the same class as a dwarf, T_{eff} must thus be lower: the drop from an M dwarf to a giant of the same class is about 300 K.

However, the effective temperatures of giants and especially supergiants are in general not known as well as those of dwarfs because we have difficulty with the theory of the distended outer atmospheres of these enormous stars. The problem is worst for the carbon stars, particularly those of type N (roughly C6–C9), with their terribly complex spectra that are often severely distorted by circumstellar clouds of cold gas and dust. We are not even sure of the class of M star that corresponds to C8 and C9. The temperature at the end of the giant and supergiant sequences may be as low as 1500 K, but the exact value is unknown.

The deep color of the M and C stars present us with another problem that we must confront. At their low temperatures of around 3000 K only a small part of their radiation spills out into the visible part of the spectrum shortward of 7000 or 8000 Ångstroms (Figure 1.11). The M stars are relatively much brighter in the infrared, and that is usually the best part of the spectrum in which to observe them. There, we see lines of many compounds, and can often detect circumstellar gas and dust lost by the stars through outflowing winds. Such a late-type star will then look dimmer than it should for its luminosity. The problem is grossly exacerbated by the molecular carbon and TiO bands.

The amount of infrared radiation relative to that in the red and yellow depends strongly upon temperature and spectral type, and thus the classical visual magnitude is a poor indicator of total radiant luminosity. For that, we use a new term, the *bolometric magnitude*, which is appropriate to *all* of the energy emitted by the star. It is found by subtracting from the visual magnitude a *bolometric correction* that depends on spectral type, or color. The correction is near zero for yellow stars like the Sun, but on the main sequence it is 1.2 magnitudes at M0, 2.3 at M5 and an astonishing 4.0 at M8, that is, for this last spectral class, the visual magnitude indicates only 2.5% of the true luminosity. For the reddest giants, we do not ever know what the values are! It is this progressive displacement of the radiation into the invisible infrared toward the latest spectral types, and the high bolometric corrections, that make the main sequence appear to plummet in visual magnitude past type M5. We run into a similar problem at the hot end of the spectral sequence, where a large portion of the energy emerges in the ultraviolet. Values of bolometric corrections are summarized for all classes in Table 12.1.

Now from all these generalities, let us now move more toward specifics, and discuss some individual stars and the various classes of behavior into which they fall.

4.4 Giants and supergiants

Just how big *are* these stars? We can estimate their dimensions from effective temperatures and luminosities by using yet another variation on the luminosity–temperature–radius relation (Section 1.15). We now solve it for radius, so that $R^2 = L/4\pi\sigma T^4$, and R becomes proportional to the square root of luminosity divided by the square of the effective temperature, where the latter is determined from spectral characteristics. If we consider L and T_{eff} relative to the Sun, we can easily find the radius of a star in solar units. The brightest of the nearby supergiants is μ Cephei, an M2 Ia supergiant with an absolute visual magnitude of -8.2 and an absolute bolometric magnitude of about -9.7, which makes it roughly 600 000 times more luminous than our 5th magnitude Sun. If we adopt temperatures of 3300 K and 5800 K for μ Cep and the Sun, we find that the supergiant has a radius 2400 times solar, or an astounding 11 astronomical units: this star would fill our planetary system out to the orbit of Saturn! Our prototype supergiants, Betelgeuse (M2 Iab) and Antares (Ml Ib) are both roughly comparable to the orbit of Jupiter. The giants, though smaller, are hardly inconsiderable. Mira, an M7 IIIe star six magnitudes brighter than the sun would reach Mars; even the smallest of the M giants if placed at the Sun might extend $\frac{1}{4}$ of the way to the orbit of Mercury and be over 10° across in our sky, double the angular separation between the Pointer stars of the Big Dipper.

The most direct way of finding stellar diameters of course is to measure both the angular size and the distance. We know that the Sun is 1.39×10^6 kilometers across because it is 32 minutes of arc wide in our sky and is 1.50×10^8 kilometers away. However, atmospheric seeing or twinkling (technically called *scintillation*) smears out the stellar images and makes it impossible to see their disks directly. Nevertheless, some of these stars are so big that we can still determine their angular diameters by using a variety of special techniques. The original such measurements were made by A. E. Michaelson at Mt. Wilson observatory in the 1920s. His method employs an *interferometer*, which measures the diameter from the way in which the light rays that come from opposite edges of the star interfere with one another (see Section 2.13). Even with this relatively primitive equipment Michaelson found that the supergiants Antares and Betelgeuse are some 0.02 seconds of arc across. A more modern version of the device, operating in Australia, employs separate telescopes and can measure diameters some ten times smaller.

The Moon provides us with another very effective method. This body moves against the stellar background at 0.55 seconds of arc per second of time, and is always passing in front of, or *occulting*, stars. We merely see how long it takes for the star to disappear behind the lunar limb. To the eye, the effect is unnoticeable: the star winks out immediately and actual measurement requires the use of a sophisticated high-speed (rapidly responding) photoelectric photometer (Section 1.11). The Moon, of course, is small, only half a degree in diameter, so the number of stars it covers in a single orbit is small. However because of the precession, or wobble, of the lunar orbital plane, we can actually observe any star within 5° of the ecliptic sometime within an 18 year period, which provides quite a large sample.

The most effective general way of measuring angular diameters is by *speckle interferometry*, which makes use of very short imaging exposures to freeze the effects of atmospheric scintillation. We will look at this method in more detail in Chapter 6. The lunar and speckle techniques have limitations of a few thousandths of a second of arc. Both are also effective in the discovery and examination of very close binary stars.

The physical sizes calculated from angular diameters and distances turn out to be reasonably in line with those found from their luminosities and temperatures. And, as pointed out above, the method provides the fundamental means for determining the effective temperatures themselves from the radii and luminosities. The problem is that the largest stars, the supergiants, those with easily measurable diameters, tend to be so far away that we do not know their distances well; and the nearby stars, with easily determinable distances (see the next Chapter) are so tiny that their angular extents are insensible. Consequently, the total number of stars for which all information is available is relatively small. In addition, the very natures of the cool giant stars cause serious problems: the gas densities are so low that it becomes difficult to define exactly what is meant by a 'surface'; moreover, many of these stars are at least partially obscured by dust of their own making. Thus it is often hard to relate the diameters found from the different methods, which hinders our theoretical interpretations of the observed spectra.

4.5 Mira stars

A fundamental feature of M (and carbon star) giants and supergiants is their instability, which makes them commonly vary both in total luminosity and temperature. The supergiants and many of the giants fall into a class of variables called 'irregular' or 'semi-regular'. Betelgeuse, for example, can be seen to change in brightness with a range of about one-half magnitude over a period of time measured in years.

The real showpieces, however, are the *Mira* or *long period*, variables (LPVs), named after the prototype *o* Ceti, which we introduced briefly in Section 4.1. Because of the interior construction of these stars, and the way in which their energy is generated by the nuclear burning of both hydrogen and helium, they physically oscillate with long, slow, fairly regular periods. Mira (which means 'the wonderful', a clear reference to its behavior) itself varies between apparent magnitudes 3 and 10, with a period of 330 days (Figure 4.7). The size of the variation is somewhat erratic, the star on rare occasions brightening nearly to first magnitude. We thus have a curious situation, in which Mira is a prominent part of its constellation for a month or two, and then is completely gone from view.

All Mira stars are giants, and can be of any class, M, N (late C) or S, usually with emission lines present; χ Cyg (S6e), which has reached third magnitude, is the most prominent example of the latter type. The subtype varies over a few tenths (Mira from M5 to M9) during the oscillation period and the class often listed in the catalogues is that at the average maximum. As they vary, the LPVs occupy a rather distinct zone in the HR diagram (Figure 4.2). Note, however, that not all stars that fall here are Miras: most, like β Andromedae (M0 III), are ordinary giants without the internal construction necessary for that kind of pulsation. We will examine the origin of the phenomenon in the last chapter. The downturn past M3 is the result of the lower temperatures, and the rapidly increasing bolometric corrections. The total bolometric luminosities actually increase toward later types: a Mira whose mean class is M8 averages about 6 times the brightness of one at M1.

The cooler, more luminous stars must also be the larger: radii range from about 100 times the Sun's at M1 to perhaps 500 at M8. The periods of the ordinary Miras, which range from about 100 to 600 days, must also correlate with mean spectral type (look at the periods of the stars whose spectra are described in Figure 4.3), as the cooler, larger stars simply take longer to oscillate.

The huge visual range of variation observed with the eye is deceptive. The pulsation causes a radius change of perhaps a factor of two, which in a complicated way in turn alters the effective temperature by a few hundred degrees. In response, the total bolometric luminosity varies by only about one magnitude. However since the eye sees only the end portion of the spectral energy distribution of these red stars, a small change

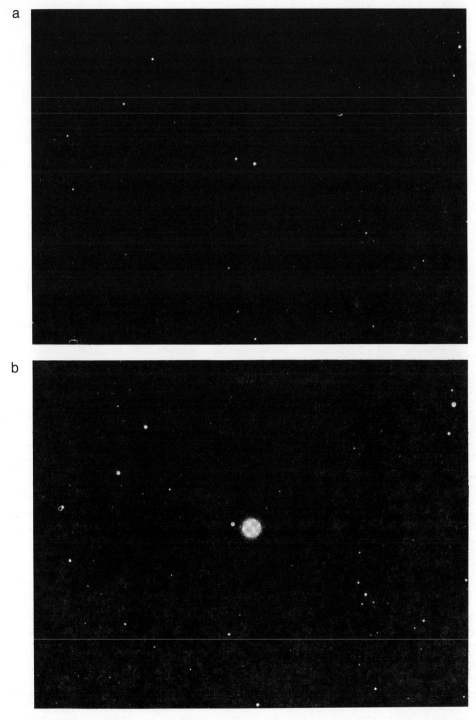

a

b

Figure 4.7. Mira in visible light at minimum (*a*) and maximum (*b*). Lowell Observatory photographs.

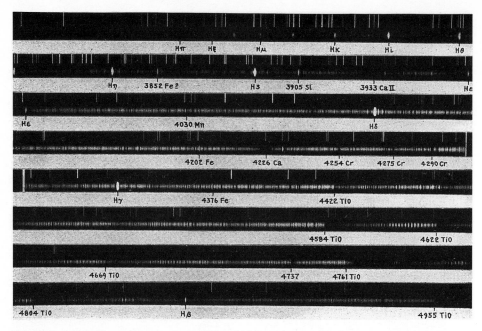

Figure 4.8. A high dispersion spectrogram of Mira, the M7e archetype of the long-period variables. Note the incredible complexity of this spectrum, which has much higher resolution than those used to evaluate spectral class; compare with Figures 3.2 and 3.5. The hydrogen lines are clearly seen to be in emission. Mt. Wilson Observatory spectrogram from *Spectra of Long-Period Variable Stars*, by P. W. Merrill, University of Chicago Press, Chicago, 1940.

in temperature can produce a large variation in the amount of visible light. The effect is greatly exaggerated by the increasing strengths of the optical TiO bands as the star cools (Figs. 4.4 and 4.8). Because the visual brightness is so dependent upon temperature, it is actually not in synchronism with the bolometric variation, which rests more on the stellar radius. For the full set of Mira stars the total average luminosity increases with advancing spectral type and decreasing temperature; but for a *single* star, the *visual* luminosity drops as the star cools in its variability cycle.

The spectrum of a Mira frequently contains hydrogen emissions that show complicated changes during the period of the variation. These features are easily seen in Figures 4.8, 3.5, and 3.2. The lines are absent at minimum light, develop strongly through maximum, and then finally disappear with the new approaching minimum. Their relative strengths are bizarre: in a normal emission source like a diffuse nebula (an excited cloud of interstellar gas surrounding a hot star), the first line of the Balmer series (Section 2.4), $H\alpha$, is always strongest, but in Miras, $H\delta$ (the $6 \to 2$ transition) dominates. The phenomenon is due to absorption of the light by various TiO bands, which shows that the emission is produced in the deeper visible strata. Moreover, Doppler shifts (Section 2.12) demonstrate that the layers responsible are moving outward. We appear to be seeing the effect of shock waves, similar to sonic booms, propagating upward in a low density atmosphere as each oscillation develops. These phenomena are almost certainly related to the ejection of mass seen to take place from these stars.

Thousands of Mira type variables are known, more than any other kind. In addition to the two already mentioned, several more can be seen at maximum with the naked eye, for example R Leonis, and R Cygni (another S star). Hundreds can easily be observed with only a small amateur telescope. There are so many, and the periods are so long, that it would be an impossible task for the world's professional astronomers to keep track of them. That work is done in excellent fashion by the American Association of Variable Star Observers (AAVSO) and the Variable Star Section of the British Astronomical Association, whose organizers compile light curves from visual estimates of magnitude made by hundreds of amateur astronomers. Other countries have similar groups. The vast body of work compiled by these organizations over the past several decades is the origin of most of our knowledge about the statistical distribution of period and amplitude, and of the changes that take place in these quantities.

4.6 Mass loss

If the primary fact of life for a giant is variability, certainly the second one is mass loss. Among the most fundamental statements that can be made about all of stellar evolution is that stars end up with much less mass than they start with. The more massive stars may lose 80% of their initial matter back into interstellar space, and even the Sun is expected eventually to reduce itself by half. This basic phenomenon has profound implications for the evolution of our Galaxy, because the stars created within it themselves become fonts of matter that will be recycled into later stellar generations. And the vast bulk of the mass loss takes place in the red giant and supergiant realms of the HR diagram, where the huge sizes of the stars make possible the required low surface gravities and escape velocities.

Optically, we can cite evidence for mass outflow from spectroscopic characteristics such as emission lines, and from absorption lines superimposed by the expanding cloud of gas upon the spectra of close binary companions. But the real demonstration comes from observations in the infrared and radio parts of the spectrum. From the former we find that the Miras are frequently surrounded by expanding clouds of dust, solid grains that have condensed out of the gaseous ejecta. We see spectroscopic features of silicates (oxygen compounds) around the M stars, and carbides (SiC) around those of class N. We believe that the carbon stars are enveloped by graphite or amorphous carbon particles of their own making. In the far infrared we can also detect the blackbody radiation of the dust itself, as it is heated to a few hundred degrees by its stellar source.

The radio spectrum shows us even stranger behavior. Here, we observe molecular line emission emanating from closely spaced rotational and vibrational states (Section 2.10) that do not differ very much in energy. The radio spectrum has opened up vast areas of research, allowing us finally to examine the chemistry of the cool gas of interstellar space. A number of simple molecules are detected in the circumstellar shells. Those associated with the oxygen-rich M-type Miras are often remarkable natural *masers*, the microwave or radio analogue to the familiar *laser*.

These words are acronyms that stand for *m*icrowave (or *l*ight) *a*mplification by the *s*timulated *e*mission of *r*adiation. In Chapter 2 we examined two atomic radiation processes, absorption and emission. An electron can absorb a photon if it has just the right amount of energy that will lift it from a lower to a higher orbit; and an electron can spontaneously jump downward, emitting a photon that has an energy equal to the energy difference between these upper and lower states. There is a third process: a passing photon that already possesses this energy can *stimulate* an electron in the upper orbit into a downward jump, resulting in two photons that go off in the same direction and with the same phase (that is, their waves are in synchrony). These two can stimulate two more, and so on, and we then see a cascade that produces a powerful monochromatic (single wavelength) beam. In a normal gas in equilibrium (that which produces blackbody radiation for example) the number of upward absorbing transitions per second between two levels always equals the number of downward spontaneous plus stimulated emissions. If, however, we can over-balance the upper state in excess of the normal relative populations that are produced by collisional excitation, and that are appropriate to the temperature, then we can generate an excess of stimulated emission. The result is an anomalously strong and sometimes powerful emission line. In the laboratory we can rather easily supply external energy to pump up the number of electrons in the pertinent upper level. In the Miras the energy source is the starlight itself.

A typical circumstellar shell is highly structured. From the cloud of an oxygen-rich M-type Mira, we sometimes observe masering lines of silicon monoxide coming from the inner part, those of water farther away, and, more commonly, very intense hydroxyl (OH) masers from the outer segment. The last, together with the powerful infrared radiation, lends the name 'OH/IR source' to such an object. The OH radiation is produced in a thin shell, and varies in concert with the star. The delay in time between the stellar and OH maxima, when divided by the speed of light, yields the diameter of the shell without our knowing the distance of the star, providing one of the few instances in astrophysics where we can obtain accurate dimensions. Typically, the OH zones are embedded in the cloud at about a thousand stellar radii from the star's surface. A measurement of angular diameter then yields the much-sought-after distance.

The clouds around some of the OH/IR sources are so thick that they literally bury the star. Sometimes all we can see of these strange celestial sources is the cool dust shell radiating at perhaps 500 K–1000 K, far too low to produce light in the optical spectrum. These stars, actually very luminous, were discovered in surveys made with sensitive infrared and radio detectors in the 1960s and 70s. Only with careful observation and analysis in the long-wave infrared, where the obscuration is low, can we detect the hidden Miras, which have periods that may extend up to 2000 days, far above that of the optically visible variety. The mass loss rates that can produce such dense shells must be enormous, and at maximum are estimated to be about 10^{-5} solar masses per year. While at first, this may seem like a very small number, remember that the star may be in this configuration for over 10^5 years, which can result in one or even several solar masses being lost to space. These stars are clearly undergoing a dramatic event in their lives: we think that they are in the process of becoming planetary nebulae, a subject that we will deal with later in this book.

We also find an analogue to this behavior among the carbon stars. While their ejecta exhibit no masers, they are much richer in molecules. Radio and infrared observations show the existence of nearly twenty different species in the archetype, a star known as IRC + 10216, including such complicated organic constructions as HC_7N and CH_3CN.

4.7 Chemical alterations

These cool stars also exhibit a variety of alterations in chemical compositions. Many of the giants seem to be in a progressively increasing state of internal circulation that brings by-products of various thermonuclear fusion reactions up to the surface. The carbon stars (Section 4.2) are clearly the result of this process, wherein this element was created earlier by the nuclear burning of helium. We also see changes in nitrogen abundances, isotope ratios, and in the numbers of the heavy atoms such as zirconium, yttrium, and lanthanum. These latter are called 's-process' elements, and are created by the successive capture of slow (hence 's'), low-energy neutrons released during thermonuclear reactions. The appearance of ZrO and YO in S stars is in part due to alterations of this sort. The proof that these elements are freshly manufactured is the widespread observation in giants of the radioactive element technetium (Section 2.2 and Figure 4.9), whose observed isotope has a half-life of only 200 000 years (the time required for a given amount of radioactive matter to decay to half its original mass), much less than the ages of the stars.

Differences in isotope ratios provide another potent example of these chemical mutations, for which the M stars are again of critical importance. The atomic spectrum of an isotope, say ^{13}C, is identical to that of ^{12}C except for a slight shift in the line wavelengths caused by the heavier nuclear mass; the larger the mass number the shorter they are. But the separations are so tiny that they are ordinarily swamped by the various

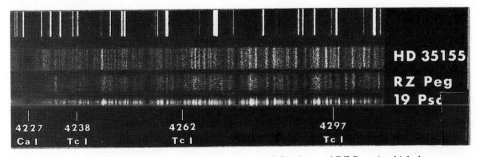

Figure 4.9. Lines of the radioactive element technetium in 19 Piscium and RZ Pegasi, which demonstrate that transmutation of atoms can take place in stars. For comparison, Tc is absent in HD 35155. Mt. Wilson and Palomar Observatory spectrograms, from an article in the *Publications of the Astronomical Society of the Pacific*, by B. F. Peery, P. C. Keenan, and I. R. Marenin.

line-broadening mechanisms, so that the separate lines cannot be seen. However, in the molecular spectra prevalent in cool stars the shifts are greatly increased and are easily visible. The CO lines then allow us to find the ratio of the abundance of ^{13}C relative to that of ^{12}C. On Earth it is about 80 to 1, but in the red giants it can be driven downward to nearly 10 to 1 through nuclear processing. This ratio, as well as others, in fact provide sensitive tests of the theories of stellar structure and evolution. The differing ratios of zirconium, derived through ZrO measurements, play the same role in S stars.

Through the various processes of mass loss, the altered mix of elements is returned to the interstellar gases out of which new stars will form. The evolution of stars, responsible for the very creation of the cool giants, in this way contributes to the evolution of the Galaxy as a whole.

4.8 The dwarfs

From the very large, we move on to the very small, with nothing in between. Henry Norris Russell, as early as 1913, recognized the profound separation of the M giants and M dwarfs. At M0 we drop through more than 6 magnitudes – over a factor of 200 in luminosity – from the faintest giants to the brightest dwarfs, whose absolute visual magnitudes are no more than about +8. In order for an M0 dwarf to be visible to the unaided eye it would have to be within about 4 parsecs (13 light-years; recall that the nearest star, α Centauri, is 1.3 pc or 4 light-years away), and none is. At M8, a typical dwarf would have to be only 0.13 pc distant. If the faintest star known on the main sequence, Van Biesbroeck 10 (VB 10, type M8 V), were in Jupiter's orbit, at opposition it would shine in our sky at only magnitude −10, ten times fainter than the full moon!

The real significance of the M dwarfs lies in their number. As we proceed down the main sequence from O to M, stellar galactic density – stars per cubic parsec – greatly increases. The O types are extremely rare, and even the Bs and the As are not too common. We see a lot of them in the naked-eye sky because they are so luminous that they are visible over great distances. Individual O stars can be seen telescopically even in galaxies millions of parsecs away. But most stars in the Universe are dwarf M; for reasons that we simply do not comprehend, nature places the vast majority into this dim category.

The masses of individual giants are relatively high, at the least equal to the Sun's, and those of the supergiants can be several times that. They have evolved from even more massive earlier type dwarfs (Section 1.17), and recall from above that they have lost some (or even most) of their matter in the process. But the masses of the as yet unevolved M dwarfs are small, at most only $\frac{1}{2}$ that of our own star, proceeding down to the theoretical limit at only 0.08 of a solar unit. The feebleness of these main sequence stars is in fact a direct result of their low masses. There is so little gravitational energy that the interior does not get very hot, and the hydrogen thermonuclear fuel fuses only slowly to produce radiation at a very low rate. Beneath the lower limit given above, the star cannot become hot enough inside to fuse hydrogen at all, and thus we arrive at the end of the main sequence. It is represented for us here by poor VB 10, our M8 dwarf of apparent magnitude 17 only 6 pc away.

The lifetime allotted to a star depends far more upon the rate of fuel consumption than it does upon the initial fuel supply (see Section 1.17). Thus the O stars burn out in a figurative instant, a million years for the earliest, which enhances their rarity. As we proceed down the main sequence the lifetimes increase. The dwarf M types are so parsimonious with their hydrogen supply that they can live for a trillion years. The longevity combined with their propensity for formation means that about 90% of all stars are of this variety. In spite of their apparent individual insignificance, these stars constitute roughly one-half the stellar mass of our entire galaxy, reason enough to devote serious effort to their study: research rendered quite difficult because of their dimness.

4.9 Flare stars

The low mass of an M dwarf confers two peripheral benefits which we will look at here and in the next section. A small, but unknown, percentage of them are variables of a type known as *flare*, or *UV Ceti stars*. This group includes the closest star of all, Proxima Centauri, the widely separated companion to α Centauri proper (see Section 1.6). These stars experience sudden, unpredictable outbursts that last but a few minutes, during which they may brighten by one or more magnitudes (Figure 4.10). The flares appear to be akin to those seen in the chromosphere of our Sun above spot groups. This layer is a low density transition region between the visible surface, the photosphere, and the extended high-temperature solar corona. All the flare stars have

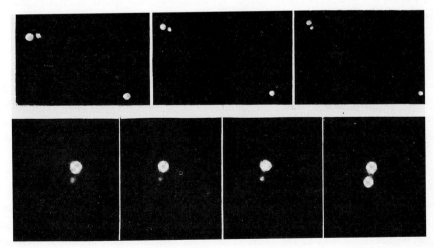

Figure 4.10. The binary star Krüger 60, in Cepheus. The upper set of photos shows the orbital motion of this close pair, spaced only about 2″ apart, over a 12 year period. From careful observation and the application of orbital theory, we can find the masses of the individual stars (Section 1.6). The lower set displays a flare on the fainter component, Krüger 60B, also known by its variable star name DO Cephei. For a brief interval the M4 V component is almost as bright as the M3 V principal member, which is usually a full magnitude more luminous. The comparison of the stars during normal light dramatically illustrates the sharp drop in absolute magnitude relative to spectral type along the dwarf sequence. Upper: Yerkes Observatory photos by E. E. Barnard; lower, Sproul Observatory photos; panel from *Burnham's Celestial Handbook*, Dover Publications, New York, 1978.

emission lines in their spectra (Proxima is a dM5e star) that suggest the existence of similar strata. We will look more closely at the details of these structures and the phenomena related to them in Chapters 5 and 6.

The eruptions on the M dwarfs are very energetic, perhaps 100 times more so than the solar variety. During an event, which may recur hours, days, or weeks after the previous one, these dim red stars become visible at short wavelengths, even in the X-ray part of the spectrum. As bright as the flares are, they would be difficult to detect and study were it not for the faintness of the normal M-dwarf background. Observations of the flares may provide further information on tenuous stellar chromospheres, on 'star spots', and on stellar activity analogous to the 11-year solar cycle, and may thus be important to our understanding of the Sun itself.

4.10 Unseen companions

The other benefit accrues more directly from the stars' low masses: they are more easily deflected by gravitational fields, and therefore are primary targets in searches for planetary systems. Recall from our earlier examination of binary stars in Section 1.6 that both members orbit a common center of mass. In the case of the Earth–Moon system it is this point that orbits the Sun on a smooth ellipse; viewed from space, the Earth would move with a slight monthly wobble about that path.

We can see the same effect in the proper motions (the movements of stars across their lines of sight: see Section 1.9) of certain stars. Proper motion is dramatically demonstrated by the movement of Barnard's Star (M5 V, Figure 4.11). Only 1.8 pc away, it is second in distance to α Cen, and holds the angular speed record of some 10 seconds of arc per year. The binary Krüger 60 in Figure 4.10 provides an example of the subject at hand. From the upper photo strip, which shows the stars in orbit around one another, it is quite obvious that the components must move in wavy lines as the center of mass of the system proceeds along a straight track. Imagine now that one of them is too faint to be seen. The oscillatory movement of the other will allow us to deduce information about the invisible object. We can estimate the mass of the visible primary star from its spectral class and the theory of stellar structure (which relates mass to luminosity). Then from the size of the deflection from the straight-line path we can infer the mass of the secondary component. The technique is widely used by astronomers to search for faint binary companions, and is a prime method for finding planets. Even a body like Jupiter, however, is light compared with a star, and the deviation it imposes upon our Sun is so small that it would be undetectable with our current technology even from α Centauri. However, if a massive planet were orbiting close to a dM star, which could have a mass only $\frac{1}{10}$ that of the Sun, and which could be 10 times as susceptible to gravitational deflection, it could possibly be found. Radial velocity measurements could also provide such a detection, as a lower mass M star would move faster in orbit, allowing easier measurement of Doppler shifts (see Section 2.12 as well as Section 5.4 ahead).

At present, we do find evidence for 'dark companions' orbiting some of the nearby stars, the most notable being 61 Cygni, whose visible members are a K5–K7 dwarf pair

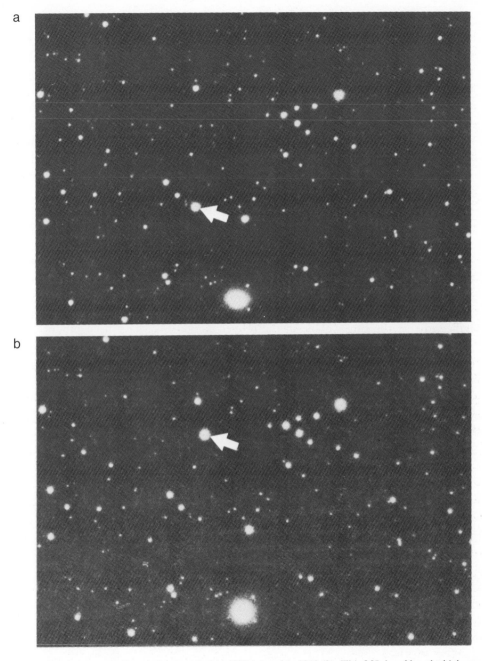

Figure 4.11. Barnard's Star, as photographed in 1937 (*a*) and in 1960 (*b*). This M5 dwarf has the highest known proper motion, and was once suspected of harboring planets similar to Jupiter. Sproul Observatory photographs, from *Burnham's Celestial Handbook*, Dover Publications, New York, 1978.

(see Figure 5.12) only 3.9 pc away. But the suspected invisible component of the trio still has a mass of about one percent that of the Sun, some 10 times that of Jupiter (with a large associated error), and although the distinction may be largely semantic, we may not really be prepared yet to call it a planet. Objects of this sort have masses that are less than the theoretical limit required to sustain thermonuclear reactions, and are currently termed 'brown dwarfs' – stars that in a sense have missed the main sequence due to their low masses, but that still may be glowing from gravitational contraction. Of the stars currently categorized as dM, only one or two are candidates for such brown status, including a highly controversial possible companion to VB 8, itself near the end of the main sequence. This particular one was inconclusively detected by speckle inter-ferometry, not by proper motion, and likely represents a false detection. No brown dwarf has as yet been confirmed; obviously we know very little about them, including how numerous they are and how they relate to the formation of planets like those in our system.

In the 1960s, astronomers at Sproul Observatory announced, after decades of study of its proper motion, the discovery of one or more planetary companions to Barnard's Star. These appeared to have masses only 2 or 3 times that of Jupiter, and we now seemed to have entered the domain of true planets. However, later work strongly indicated that the tiny wobble in the proper motion may have been produced by uncertainties associated with the telescope, illustrating the immense difficulty of the problem. So far, no extra-solar Jupiter has been found. But we are hoping that the next generations of telescopes and detectors, observing from Earth-orbit, may let us to find such bodies, opening up a brand new field of very exciting research, and allowing us to examine new 'solar systems' that we already well suspect are out there (we will look at some of the evidence later). And the dM stars will be among the first to be examined.

Figure 5.1. The great orange star, Arcturus, K2 IIIp, fourth brightest to the eye. The first spectroscopists were just barely able to record its spectrum on the primitive photographic plates available. With advanced digital scanners, we are now able to examine the spectra of some of the faintest stars in the photograph, near magnitude 20. ©1960 National Geographic–Palomar Observatory Sky Survey. Reproduced by permission of the California Institute of Technology.

5 The K stars: orange giants and brighter dwarfs

Three beautiful gems vie for the title of brightest in the northern hemisphere. Blue white Vega, class A0 V, marks the summer skies for us, only 0.05 magnitudes brighter than yellow Capella (G5 III + G0 III double), of winter prominence. As one sets in the northwest the other ascends the northeast. Positioned between them, and just slightly the brightest of the three, is our archetypal K star Arcturus (K1 III, Figure 5.1 and 5.2). When northerners see its prominent orange light low in the east in early evening, they know that spring will soon follow.

Figure 5.1.

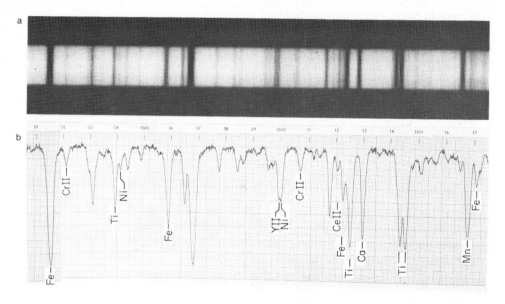

Figure 5.2. (*a*) A detail of the spectrum of Arcturus between 5501 Å and 5518 Å, taken at extremely high dispersion. This star is so bright that it has been used to make a standard atlas of lines. Note the extraordinary number of features present even in this short 17 Ångstrom segment. (*b*) shows a tracing of the spectrum, in which intensity is graphed as a function of wavelength. Several atomic lines are chemically identified. Mt. Wilson and Las Campanas Observatories, Carnegie Institution of Washington, spectrogram, from *A Spectral Atlas of Arcturus*, by R. F. Griffin, published by the Cambridge Philosophical Society, 1968.

Several other familiar stars, easily given away by their notable color, fall into this class: Pollux (K0 III), contrasting smartly with white Castor (Al V + A2 V: remember it is a visually observed double); the top front bowl stars of both the Big and Little Dippers, Dubhe (α UMa, K0 III) and Kochab (β UMi, K4 III); and of course the Bull's eye, Aldebaran (K5 III). The last is the latest and reddest in color, and for Father Secchi represented a transition between types II (the yellow-orange stars) and III (the red kind with fluted spectra).

Only 20 pc away, Aldebaran is coincidentally projected against the doubly distant Hyades star cluster (Figure 5.3), itself a notable cache of K giants. The prominent stars of the 'vee' of Taurus' head, θ^1, γ, δ^1 (Figure 5.4), and ε Tau are all of class K0 III. More distant sets are distinctive in the Praesepe or Beehive Cluster (Figure 5.5) in Cancer, and in many similar stellar assemblies.

Although the best known of class K are giants, we should not slight the dwarfs. Unlike type M, some of these are visible to the unaided eye. The most famed is surely 61 Cygni (see Figure 5.12 ahead), the first star to have its distance actually measured. This K5–K7 double shines at apparent magnitude +4.8, easily seen in a dark sky 8° southeast of Deneb. The visually brightest on the K main sequence is the fainter member of the α Centauri pair (K1 V), which because of its proximity to us shines at apparent magnitude 1.3; if the primary component (G2 V, like the sun) were not present, α Cen B, at only 1.3 pc, would still be the 20th brightest star in our sky.

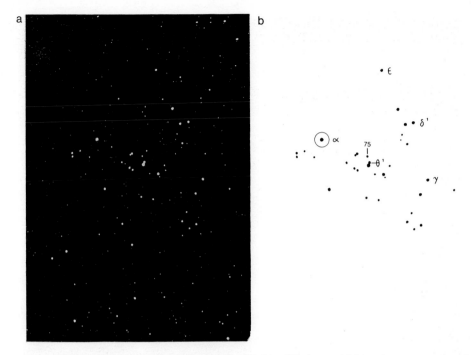

Figure 5.3. (a) The Hyades, with its prominent membership of K giants, which are the stars named on the chart (b). The brightest star is Aldebaran, K5 III; it and 75 Tau do not belong to the cluster. Another cluster, NGC 1647, 10 times farther away, is at the upper left. Lowell Observatory photograph.

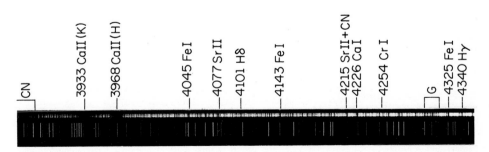

Figure 5.4. A spectrogram of δ^1 Tauri, K0 III, taken at moderate dispersion with the Kitt Peak 84-inch telescope and coudé spectrograph. Note the great complexity of the spectrum, which exhibits many molecular and metallic lines and very strong H and K of Ca II. One set of comparison lines runs along the bottom. National Optical Astronomy Observatories (Kitt Peak) spectrogram, courtesy of K. M. Yoss.

a

b

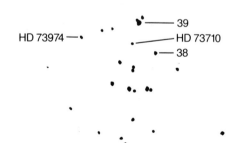

Figure 5.5. (*a*) The Praesepe, or Beehive Cluster (Messier 44) in Cancer. This familiar assembly, about 4 times farther away than the Hyades, contains four bright K giants indicated by name on the chart (*b*). These stars appear somewhat dimmer on this blue-sensitive plate than they would to the eye. Delta (lower left corner) and η (upper right corner) Cancri are also K giants. ©1960 National Geographic–Palomar Observatory Sky Survey. Reproduced by permission of the California Institute of Technology.

Although still very distinct, the giants and dwarfs are not separated by quite the gulf that is apparent for the M stars. The K giants are fainter and smaller than their M counterparts (Figure 5.6), and the dwarfs brighter and larger, as luminosity class III begins to merge toward type V. Arcturus, 25 times the size of the sun, is only half the diameter of the M0 giant β And, and 61 Cyg A, with a radius 0.7 solar, is 3 times bigger than Barnard's star. But we still see enormous, and sometimes very strange, supergiants, and among the K types we encounter our first subgiants and white dwarfs. All considered, we now examine a most interesting set of stars.

5.1 Classification

Many spectral features attest to the K nature, and to atmospheres significantly warmer than those found in the M stars. As we progress toward earlier subtypes, the TiO bands finally disappear at about K5, and the hydrogen absorption lines make their appearance among all luminosity classes. The powerful g line of neutral calcium steadily weakens as ionization becomes important, and H and K of Ca II emerge clearly out of the developing violet part of the spectrum (Figure 5.7). The very strong D line of Na I (actually a closely spaced doublet) in the yellow behaves similarly to its neutral calcium counterpart. Ratios of several other metallic lines are also quite useful for classification. The higher temperatures, which cause the break-up of most molecules, allow only the more tightly bound ones to survive: the G band of CH increases in strength well into class G, and in the giants, the optical CN and infrared CO bands are prominent as well.

The luminosities of the K stars are as easily found (Figure 5.8). The hydrogen and neutral calcium lines behave much as they do in M stars, especially for the later subtypes (respectively exhibiting negative and positive luminosity effects: see Section 4.1). Various ratios of metallic absorptions, particularly ionized strontium and titanium to neutral iron (which show a positive effect), are also good indicators. The most obvious signatures are probably molecular bands of CN that begin abruptly at 4215Å and 3888Å and extend into the blue. The former is known as the 'CN break'. CN greatly strengthens from the main sequence toward higher luminosity as the lower atmospheric pressures promote its formation. However common abundance variations render it tricky to use.

The use of any single spectroscopic criterion can lead the astronomer into serious error. For example, the hydrogen lines in types K and M strengthen with both increasing temperature and luminosity. On the basis of these lines alone, a giant would be assigned a spectral type that might be as much as a full class too early relative to one implied by the metallic or molecular lines. Accurate classification can be done only by taking *all* defined criteria into account. For a newly observed star of unknown variety, we would first use features sensitive *mostly* to temperature in order to find its Draper class, such as the hydrogen lines, or TiO bands, much as outlined in Section 4.1. Then we would find the luminosity from those criteria responsive primarily to density, like CN at type K. Once we know the star is a giant, dwarf, or supergiant, we can then refine

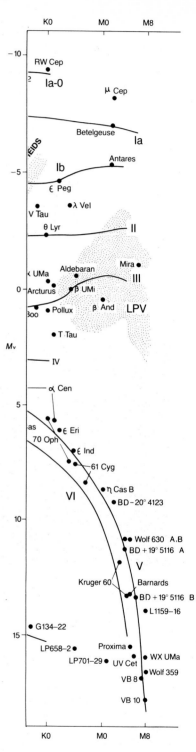

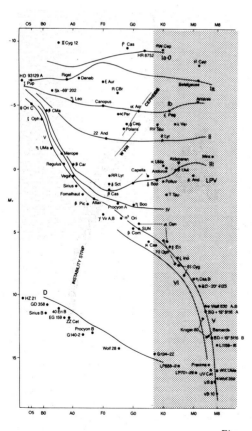

Figure 5.6.

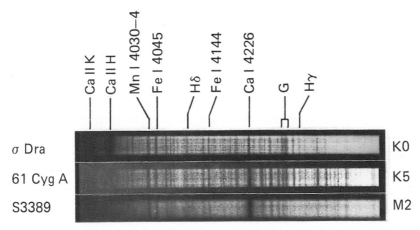

Figure 5.7. Spectral variations through the K main sequence and their comparison with an early M dwarf. Note the weakening of Ca I and the strengthening of Ca II as the temperature climbs from M to K. A photo of 61 Cyg is shown in Figure 5.12 below. From *An Atlas of Stellar Spectra*, W. W. Morgan, P. C. Keenan, and E. Kellman, University of Chicago Press, Chicago, 1943.

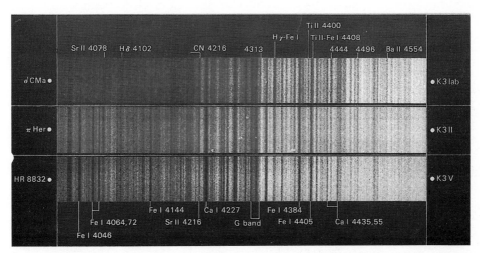

Figure 5.8. Luminosity changes among three early K stars. Note the strengthened cyanogen (CN) absorption toward the supergiants, and the obvious positive luminosity effect at Hγ. The weakening of the Ca I g-line toward higher luminosity, so obvious in late K and early M stars, has disappeared. If carbon and nitrogen are deficient in a star, the CN break at 4215 Å weakens and the astronomer could be fooled into classifying a giant as a dwarf. From *An Atlas of Representative Stellar Spectra*, by Y. Yamashita, K. Nariai, and Y. Norimoto, University of Tokyo Press, Tokyo; John Wiley & Sons, New York, 1978.

Figure 5.6. The HR diagram for classes K and M. The domain of the long period variables extends partially into class K, and some of the supergiants are irregular variables; otherwise, stars of this class tend to be quite stable. Although μ Cephei has a fainter absolute visual magnitude than RW Cep, it is actually somewhat the brighter bolometrically, that is, in terms of total luminosity. We see here the first of the subgiants and, down at the bottom, the faintest and coolest of the white dwarfs. The latter actually extend a bit (and uncertainly) into the temperature realm of class M. See Figures 3.8 and 12.7 for credits.

its temperature class, and so on. The artful practitioner of the science does all this in an instant; it is a learned skill, not unlike recognition of species of birds and trees.

Note that spectral classification so far as we have described it is *qualitative*, not *quantitative*: we have actually measured nothing, but only have determined the temperature and luminosity of a star from the visual appearance of the spectrum. The beginning spectroscopist does this by the careful comparison of a stellar spectrum with a set of preselected standards, those of the MKK Atlas, for example. With improving skill, the standards become embedded in memory, and the typing of stars can proceed with impressive speed. There is a very important place for quantitative methods, whereby we classify according to measurements of the actual amount of energy extracted from the spectrum by various sets of lines, and we will touch upon it occasionally in this volume. It can be a laborious, often slow procedure, although speed is improving with the progressive development of computer classification techniques. Visual classification, however, is at present still needed in order to survey in a reasonable amount of time the vast numbers of stars readily accessible to us. It was the only means by which Annie Cannon and her followers could classify over one-third of a *million* stars in the Henry Draper Catalog, and the principle is still relevant today, as astronomers at the University of Michigan proceed with the equally vast MKK reclassification.

5.2 Spectroscopic distances

If you wished to determine the absolute radiant luminosity of a star, you would measure the brightness at Earth (in terms of the amount of energy passing us per square centimeter every second), and then multiply by the surface area of a sphere whose radius equals the stellar distance, all of course in appropriate, consistent physical units. Or, more simply, we express the luminosity astronomically as absolute magnitude, M, and calculate it from apparent magnitude, m, through distance (D) using the magnitude equation, $M = m + 5 - 5 \log D$ (Section 1.13). Of course we use the modern B or V values for m.

The beauty of the MKK classification scheme is that it allows us to determine the luminosity directly from the spectrum, without knowing how far away the star is. We can then turn the relationship around, and find the distance of the star, and whatever stellar system contains it, by comparing the observed, apparent brightness with the intrinsic value. Log D, then, is just $0.2(m - M + 5)$. The difference $y = m - M$, called the *distance modulus*, is fundamental to stellar astronomy and is a common substitute. We can easily follow the procedure, using our HR diagram. From its spectrum, we find that 39 Cancri, one of the Praesepe stars, is of class K0 III, for which the visual absolute magnitude (M_v) should be about $+0.5$. With its apparent visual magnitude, V, of 6.4, we calculate that $y = 5.9$ and $\log D = 2.2$, so that the Beehive is about 150 pc away. We may immediately see a problem though, in that giant stars have a spread about the average; 39 Cnc *might* be as bright as α UMa ($M_v = -0.3$) which would place the cluster over 50% farther away. In spite of this inherent error (which can be decreased through quantitative classification techniques), the distances so derived are frequently quite adequate for astrophysical purposes. In a cluster, at least, we can use several stars or even the whole main sequence in order to improve the precision. Somehow, however, we must first know the typical absolute magnitudes for the different types of stars, that is, we must calibrate the diagram and the various luminosity classes against M_v with stars whose distances are known in advance. We will look at that fundamental problem later in this chapter, when we examine the dwarfs.

5.3 K lines and chromospheres

The late-type stars, G, K, and M, have chromospheres, low-density regions of gas that overlay the higher density stellar photospheres where the continuum and the bulk of the absorption lines are formed. The term comes from the distinct red color of the Sun's, which is only visible during an eclipse or with specialized instrumentation. We first encountered this concept in association with the dwarf M flare stars (Section 4.9).

Chromospheres can be detected by detailed high resolution spectroscopy of the strongest lines. The most frequently used are the Fraunhofer H and K pair of ionized calcium; the latter is best, since H is blended with a hydrogen line. (Remember not to confuse 'Fraunhofer H' with the chemical symbol 'H'.) Since the calcium K absorption line of the photosphere is produced by an upward electron transition from the lowest energy state, nearly every ion is capable of producing it, which causes it to attain astounding strength and be tens of times broader than the strongest of the iron lines (see Section 2.11). This main absorption feature, which we now call K1, is the one commonly used for determining spectral class.

The star's chromosphere, however, acts differently. It is a low density, hot gas, and it generates *emission* lines (Section 2.6). If we look with enough detail, we will see a narrow bright K line, named K2, centered on the very broad absorption. However, even the gas of the chromosphere can become fairly opaque, and it absorbs some of its own radiation, which results in an even narrower absorption, termed K3, in the center of the emission. We say that the emission feature is self-reversed (or self-absorbed), and we will encounter the phenomenon again among the B stars. The ultraviolet Mg II lines at $\lambda 2796$ and $\lambda 2802$ (called h and k by analogy with their Ca II counterparts and which are visible only from satellite-borne telescopes) behave similarly.

The K-line chromospheric emission provides a superb example of quantitative spectral classification. In 1956, O. C. Wilson and V. Bappu, working at Mt. Wilson, discovered that the width of K2 correlates very well with absolute visual magnitude for classes G, K, and M, from dwarfs through supergiants over a range of more than 15 magnitudes (Figure 5.9). The cause of the phenomenon is still not clearly understood: it involves the origin of the chromosphere itself, and mass motions that generate energy. But our ignorance need not stop us from using the method to determine the luminosities and distances of stars with considerably greater precision than is possible through only simple MKK classification. Although the observations require long exposures the Wilson–Bappu effect has provided us with an excellent, now standard, distance indicator in astronomy. The major problem is again calibration – we must first know independently how far away some of the stars are. But that in part again involves the K dwarfs, so let us first examine some further topics that engage the giants.

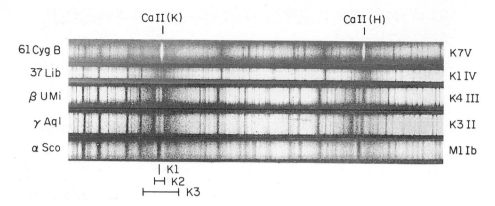

Figure 5.9. The region of the ionized calcium H and K lines in four stars of class K and one of early M. A self-reversed line, that looks like a double emission, is centered in a broader absorption. As the luminosity increases, so does the total width of the emission. This is the *Wilson–Bappu effect*, which is so useful as a luminosity, and hence distance, indicator. In the dwarf, the central absorptions (the K3 and H3 components) cannot be detected at this dispersion. Mt. Wilson and Palomar Observatory spectrograms, from an article in the Astrophysical Journal by O. C. Wilson and V. Bappu.

5.4 Eclipsing supergiants

The K stars provide us with a prototype set of bizarre eclipsing binaries that allow us to examine stellar atmospheres and chromospheres in considerable detail. Some of the individual events are even visible to the naked eye. If the geometry of a binary orbit is just right, one star passes in front of the other, and the total light of the system drops for a time. Eclipsing binary stars are generally common: there are almost as many known as Mira variables. The most famous example is Algol (β Persei), which diminishes in brightness by a full magnitude every 2.9 days as a bright small B star passes behind a G8 giant.

Stars of this kind are among the most important in astronomy, as analyses of the eclipses allow us to determine numerous stellar parameters. For example from the time it takes one star to cover another we can derive the stellar diameters: another method to add to those discussed in Section 4.4. From the drop in brightness of the system, as each component in turn hides the other, we can calculate surface brightnesses, which we can use to check our estimates of temperature.

More importantly, the eclipsing systems generate much of our knowledge about stellar masses. Most visual binaries – the kind that can be separated by eye – are so far apart and have periods so long that we cannot construct an accurate orbit. If the components are closer together we cannot split them visually, but then their orbital velocities are sufficiently high to produce Doppler shifts (Section 2.12) in their spectra (which under ideal conditions will be a combination of the two stars); see Figure 5.10.

If we know orbital velocities, we can deduce the mass ratio and the size of the orbital semi-major axes, and from Kepler's third law and our knowledge of the center of mass (Section 1.6) find the individual masses. Unfortunately, these spectroscopic

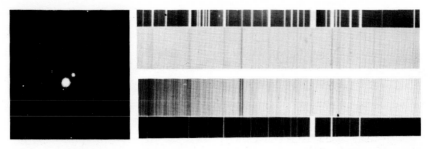

Figure 5.10. Mizar and the spectrum of one of its components. The naked-eye view of Mizar and Alcor, its distant companion, is shown in Figure 1.3. A telescopic view of Mizar alone, displayed on the left, shows it to be double with a component separation of 14 seconds, of arc. Both of these are spectroscopic binaries each with two visually inseparable components. Spectra of the brighter are shown on the right. At the top the two stars are orbiting so as to move across the line of sight. Consequently, there are no relative Doppler shifts and the spectra of both stars merge into one. At the bottom, one star is now receding in orbit, the other approaching, with the resultant blueward and redward Doppler shifts, so that each spectrum becomes visible. The Doppler shifts provide orbital velocities, which in turn give us information on masses. Photo from the Lowell Observatory; spectrogram from the Yerkes Observatory; panel from *Burnham's Celestial Handbook*, Dover Publications, New York, 1978.

binaries are likely to have their orbital planes inclined to the line of sight, so that all we can find are lower limits to both the velocities and the masses. But if the system eclipses, we know that the orbital plane must lie on edge, or nearly so; from the light curve (the way in which magnitude changes with time) we can derive the actual orbital inclination, and consequently, accurate masses.

The archetype of the set of eclipsing systems that we consider here is ζ Aurigae, the faintest of 'The Kids' of Cappella (see Figure 6.2), normally at magnitude 3.80. Every 972 days, the B8 V component passes behind a K4 II giant, and the visual luminosity drops by a barely noticeable 0.2 magnitudes. The 2.6 year period and the 38 day duration of the eclipse attest to the great proportions of the system. For about 2 weeks before and after total eclipse, we can see the extended atmosphere of the K star imposing itself upon the spectrum of the B component, which provides us with a marvelous natural laboratory for studying the structure of, and the motions of gas within, the supergiant's chromospheric envelope. The observations tell us, for example, that this star produces a wind through which it loses about 2×10^{-8} solar masses per year.

Other naked-eye stars that behave similarly are 31 and 32 (o^1 and o^2) Cygni, K2 II + B3 V and K3 Ib + B3 V binaries that are only 1° apart. If we extend to either side of class K, we also find U Cephei (G8 I + B8) and VV Cephei (M Ia + B). Both are telescopic objects, but the variations are much more easily visible to the eye; U is great fun to watch, dropping by 2.3 magnitudes every 2.5 days. Much more patience is required for VV: every 20.2 years it dims by 0.8 magnitudes for an amazing 1.2 years while the hot component hides behind the enormous cool one, which like μ Cep is of solar system proportions. Even stranger is ε Aurigae, curiously but coincidentally the brightest of the 'Kids' only 3° from ζ. The primary component is an F0 supergiant that appears to be eclipsed periodically by a ring of dust surrounding a blue companion, which itself may be a close binary. The event occurs every 27 years and lasts for 714 days, with a quite visible amplitude of 0.8 magnitudes. For those with the patience, the next eclipses of VV Cep and ε Aur will begin in 1997 and 2009, respectively.

While we are on the subject of variables, a comment on the comparative intrinsic stability of type K is in order. We see a few of the irregular variety among the supergiants (on our HR diagram, note RW Cephei, which has an erratic amplitude of about 2 magnitudes), the Mira domain extends up to roughly K5, and there are a few flare-type stars in complexes such as the Orion Nebula, but that is about all. True pulsating variables in this class are relatively rare compared to the number found among type M.

5.5 Composition variations

We now use our K stars to explore additional complexity within the 'third dimension' of spectral typing. At class M, we were forced to branch the giant spectral sequence into classes S and N (later called C6 to C9) to account for stars with carbon overabundances caused by the mixing of surface gases with the interior (Section 4.7). As we saw in Chapter 3 (see Table 3.4), the N stars continue into warmer temperatures as class R: K0 to M0 runs parallel to the sequence R2 to about R8, or in other terms, C2 to C5. We can readily recognize the R giants by the bands of C_2 in their spectra.

The S stars, which have roughly similar C and O abundances as well as enhanced quantities of s-process elements (Section 4.7), yttrium, zirconium, and the like, appeared once to extend into the K (and late G) temperature zone as the *barium stars*. These are easily recognizable by the Ba II line at $\lambda4554$, and also show strong absorptions caused by strontium, particularly $\lambda4077$ and $\lambda4216$ Sr II. In addition, we find a few 'mild barium stars' that are higher-temperature analogues to type MS (those between classes M and S). However, since there is strong evidence that the barium stars are binaries, the connection with class S has become uncertain. What appears now to be more likely is that one member of the binary has evolved and expanded, and in the process has passed some of its enriched mass onto the other component, showering it with peculiar elements. We will encounter other, more certain examples of mass transfer in close binaries later (Section 11.4).

So, far we have looked only at composition enrichments, but underabundances of various elements are of equal significance. The CN break, we noted, is very sensitive to luminosity. It is very strong in giants, and almost absent in dwarfs. For a time it was widely used to establish absolute magnitudes, but spectroscopists later found that some stars that were classified as giants by all other criteria were being erroneously called dwarfs on the basis of CN alone; that is, the 4215 Å break was too weak for the luminosity class. This phenomenon is clearly caused by carbon and nitrogen abundances that are well *below* solar, which shows that they belong to the metal-deficient Population II variety of stars that inhabit the galactic halo (Section 1.8). The difference between the observed CN strength and that for stars of solar abundance of a given luminosity is called the *CN anomaly*. The effect extends into late G as well, and related deficiencies are regularly noted within class M. Since the K giants are so bright, we can see them for great distances (easily determined through spectroscopic luminosities) from the plane of the Milky Way. Thus these stars provide us with a way of examining the carbon and nitrogen abundances within Population II, and by inference the metal composition of the early Galaxy, before the formation of the galactic disk. (We will examine the broader aspects of the lower metal content of the halo at a later time.)

The CN anomalies can also be positive, implying overabundances of C and N; if they are high enough the C_2 lines appear and the stars are called class R. The evolution-induced overabundances and the galactic underabundances can in practice be difficult to separate, producing limitations on the studies of either one. We also find odd combinations of both abundance enhancements and deficiencies. For example the *CH stars*, metal-deficient but with strong CH, s-process, and C_2 features, may be something of a population II analogue to the more prevalent carbon stars of the disk.

5.6 Toward lower luminosity

As we proceed downward from the K giants on our way to the dwarfs, we do not encounter the great empty gulf between the two that is so prominent among the M stars. Along the way we find a smattering of *subgiants* (class IV), stars well above the main sequence, but clearly not of giant status. These may have diameters perhaps 4 or 5 times the Sun's, and appear to be evolving from the main sequence upward into true luminosity class III.

In this region above the main sequence we also encounter the bulk of the *T Tauri* stars, which are also fairly common in classes G and M. These odd stars exhibit emission lines, and vary irregularly over a range of roughly one magnitude. They are grouped into what are sometimes *T Associations*, and are found in conjunction with clouds of interstellar matter. We believe that they are newly formed, and may be the youngest stars known. We will examine them again in Chapter 12.

Finally, roughly between absolute magnitudes 6 and 8 we arrive at the K dwarfs. While not so numerous in the Galaxy as their M counterparts, there are still plenty of them, perhaps $\frac{1}{6}$ the number found in the coolest class. But their masses are larger, between roughly 0.5 to 0.7 times solar, which accounts for their higher main sequence luminosities.

5.7 Parallaxes and stellar distances

Generally, these dwarfs are an undistinguished lot, but since the star with the first known distance is among their company, we have an opportunity to examine such measurements, and the calibration of the spectroscopic absolute magnitude scale promised earlier. From the time of Copernicus, and the sure knowledge that the Earth moved about the Sun, astronomers had searched for the annual parallax shift caused by our observing the nearby stars from different locations in the Earth's orbit (Figure 5.11). The great distances involved frustrated the early attempts, although one, by James Bradley, led to the discovery of the much larger aberration of starlight (the

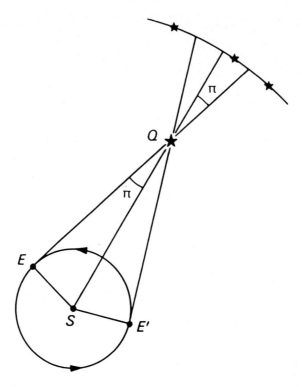

Figure 5.11. Annual parallax. The nearby star will appear to shift back and forth through a small angle as the Earth goes about the Sun. The parallax π, measured in seconds of arc, is half the total deflection. From *Principles of Astronomy, A Short Version*, 2nd edn., by S. P. Wyatt and J. B. Kaler, Allyn and Bacon, Boston, 1981.

angular displacement of a star caused by the velocity of the Earth relative to the finite speed of light) in the early 1700s.

Not until 1838 did anyone succeed, when Friedrich Bessel announced the discovery of the slight semi-annual shift of $\frac{2}{3}$ of a second of arc for the K5–K7 double 61 Cygni (Figure 5.12). He chose this star for study because at the time it had the highest known proper motion and it was rightly thought to be one of the nearest to us. Recall here that 61 Cyg is also noted for its possible submassive unseen companion, inferred from the binary's wobbly path (Section 4.10). Bessel's early measurement is quite close to the modern value. The *parallax* (π) of a star is defined as one-half of the total semi-annual shift, and that of 61 Cyg is 0.29 seconds of arc. The distance in parsecs (Section 1.4) is defined as $1/\pi$, and thus the star is 3.4 pc away. One parsec is the distance at which the angular radius of the Earth's orbit would be 1 second across, and is equal to 3.26 light years.

Somewhere around 10 000 stars have by now been subject to parallax measurement, although a great many of these are so far away that the derived values are essentially zero. At current levels of precision, we cannot accurately determine distances in this way much beyond about 50 pc, although new techniques will soon

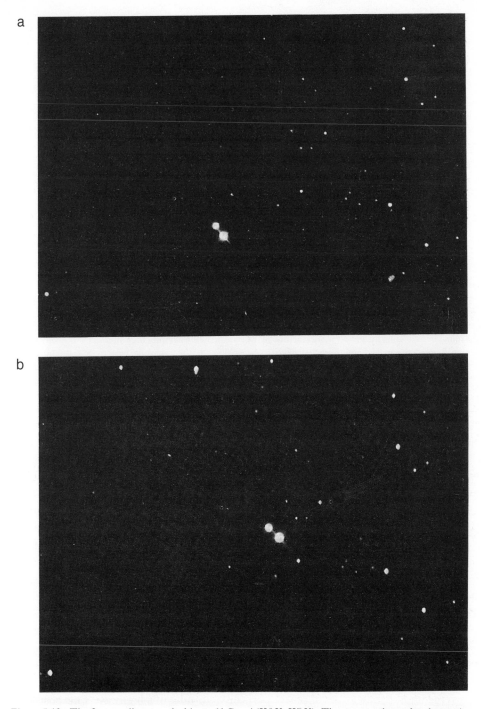

Figure 5.12. The first parallax star, the binary 61 Cygni (K5 V, K7 V). The star was chosen for observation by Friedrich Bessel because its high proper motion, easily seen here, suggested proximity. There may also be a low mass invisible companion. The two photographs were taken at the Lowell Observatory in 1916 (*a*) and 1948 (*b*).

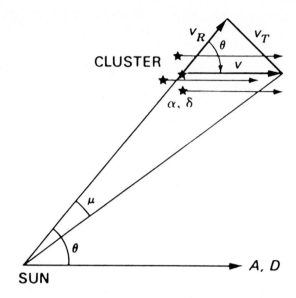

Figure 5.13. The stars in the cluster, now at right ascension and declination α,δ, will move off to the convergent point, or radiant, A,D at some time in the very distant future. The position A,D on the sky can be found by extending the proper motion arrows of the stars in the cluster until they intersect at the convergent point. Because the angle θ and the radial velocity, v_R, are known, the transverse velocity, v_T, can be found. From v_T and μ (the proper motion), we can deduce the distance of the cluster. From *Principles of Astronomy, A Short Version*, 2nd edn., by S. P. Wyatt and J. B. Kaler, Allyn and Bacon, Boston, 1981.

allow us to go much farther. This radius includes a few stars of the middle main sequence, like Sirius and Vega, and a number of prominent giants such as Arcturus, Pollux, and β Andromedae, but none on the luminous upper main sequence, and no supergiants; the vast majority of parallax stars are the dim K and M dwarfs. We can thus use these measurements to establish absolute magnitudes associated with luminosity classes V and III, and to calibrate the dim dwarfs for quantitative schemes such as the Wilson–Bappu effect (Section 5.3). But the very luminous stars, those that provide us with beacons that bridge the greatest distances, escape: they are just so rare that there happen to be none nearby.

For these brightest stars, we must do something else, and we return again to the Hyades, so prominently mentioned at the start of this Chapter. The cluster is so close that we can perform a very clever trick to obtain a highly precise distance (Figure 5.13). We know that all the cluster stars are gravitationally bound together, and that except for very slow orbital motions, their true courses are parallel to one another. But when we plot their observed proper motion directions on a sky atlas, we see that they appear to converge toward a point in space, a clear effect of simple perspective much like the convergence of parallel rows of power poles on both sides of a highway. This position, called the *radiant*, tells us where in the sky the cluster will ultimately be seen after millions of years have passed. Because we have an integral *group* of stars, we can find the angle between where they are seen now, and where they are headed. With this knowledge and the observed radial velocity (the speed along the line of sight computed

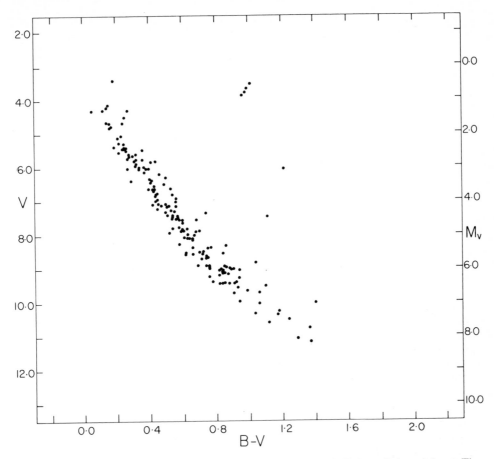

Figure 5.14. An HR diagram of the Hyades, plotted against color index (B–V) instead of spectral type. The B–V values for A0 and K0 dwarfs are 0 and 0.8 respectively. See Figure 3.10 for a more complete conversion. The lack of upper main sequence stars and the existence of a prominent giant branch are typical of older clusters, in which the early type stars have long since burned away. This plot is first combined with parallax stars, for which absolute magnitudes are known, to provide a basis for the distance determination of younger clusters, and for the absolute magnitude calibration of the brightest stars. The four K giants are identified in Figure 5.2. From *An Atlas of Open Cluster Colour–Magnitude Diagrams*, by G. L. Hagen, Publ. of David Dunlap Observatory, Toronto, 1970.

from the Doppler shifts of the spectrum lines: see Sections 1.9 and 2.12), we can compute the transverse velocity, the speed *across* the line of sight. But the proper motion – the annual angular displacement – depends upon transverse velocity and distance. With the former known, we can find the latter. For the Hyades it is 45 parsecs, known to an extraordinary precision of two percent. No other cluster, including Coma Berenices, the Pleiades, or Praesepe, is close enough to provide sufficiently large and accurate proper motions.

With the Hyades' distance known, we can establish its HR diagram (Figure 5.14) and the absolute magnitudes of its stars (including its prominent K giants). We can also add into it the many nearby parallax stars, and use the combined diagram as a standard.

Unfortunately, the Hyades is a relatively old cluster, and its once-luminous upper main sequence has long since burned away; neither O and B main sequence stars, nor any supergiants are present. Remember that the more massive a star, the shorter its lifetime (Section 1.17). But now we start working on the spectroscopy of other clusters. We establish a relative HR diagram, apparent magnitude against spectral type (or color), for, say, the Pleiades, a younger cluster with a main sequence that extends to the much more massive B stars. We next fit the two plots to one another by lining up the lower axes, and by sliding one vertically until the main sequences match. By comparing the two vertical axes, which show both apparent and absolute magnitudes next to one another, we can then immediately find the Pleiades' distance modulus, $m - M$, and of course the distance itself. This procedure now pushes our absolute main sequence up into type B. We do the same for even younger systems that contain the great O stars and genuine supergiants such as the Double Cluster (h and χ) in Perseus. With enough clusters, and sufficient labor, we can go most of the way towards finding the absolute magnitudes of the various stellar types, and determining the necessary calibrations for a variety of additional distance methods. Such a compilation of cluster diagrams is presented in Chapter 12, where we examine the evolution of the stars (see Figure 12.11).

5.8 Subdwarfs and white dwarfs

Finally, let us look briefly at the stars below the main sequence. Roughly one magnitude down, we find the subdwarfs. These are metal-poor, and bear the same chemical relation to the dwarfs as the CN-deficient stars do to the normal K giants. Actually, as we saw in Section 3.5, the subdwarfs are not too *faint*, but too *early* (or blue) by several subtypes. The lower metal content weakens the classification lines, which produces an earlier class. Moreover, the partial removal of the metal lines allows more blue light to escape, and the decreased opacity of the gas caused by fewer metallic atoms generates a smaller star and higher temperature. We see the subdwarfs among type M as well.

After a great drop of several more magnitudes, we introduce our first *white dwarfs*. The name derives from the first of the breed to be discovered, which are indeed white in color. These are truly tiny stars of terrestrial dimensions and enormous densities that have developed out of the giants (Section 1.17). They were originally classified DO (or wdO) through DM on the basis of their spectral appearance, in parallel with the main sequence. But their compositions are so strange that ordinary spectroscopic terminology loses its original meaning. The 'DK' star (LP658-2) shown earlier on the HR diagram (Figure 5.4) is plotted on the basis of its temperature (4300 K) and the main sequence temperature scale, not on the presence, or lack of, absorption features. The 'DM' star (LP701-29) also plotted is the coolest known, with a temperature of only 3750 K. For now, only note the existence of these objects; a full discussion, including an explanation of their spectral peculiarities, will be deferred until we look at the A and B stars. Before that however, we will proceed to the most familiar kind of all, the next great group, the stars like the Sun.

6 *Our sun and its cousins: the G stars*

No stars are better – and at the same time more poorly – understood than those of class G. That cryptic statement has its roots in our observations of the yellow G2 V star that lights and warms the day, the Sun (Figure 6.1). The amount of data gathered on that body over the half-century history of modern astrophysics is truly astounding. We can examine the spectrum over tiny areas of its surface, and see motions and changes with time and position in exquisite detail. Consequently, we know our own star, and by analogy class G in general, extremely well. But we have such a vast mass of information that no comprehensive theory can yet interpret it, and in that sense, the Sun remains a considerable mystery. Many solar phenomena have no real explanations at all at the present time. If you do not see an event – and there must be many strange ones awaiting our discovery on, say, F or K stars that will equally mystify us – you do not have to explain it.

We will return to this remarkable body shortly, but first let us look at some other familiar G stars, and at the properties that define this type. By the kind of random coincidences that abound in astronomy, the brightest component of our nearest neighbor, the binary α Centauri, is also a G2 dwarf, almost identical to the Sun. Curiously, this system has provided an example for all the cool classes we have encountered so far: the secondary is K1, and the third member, proxima, is an M5 flare star. Capella (Figure 6.2), the sixth brightest in the sky, is a close double that consists of G0 and G5 giants (Fig. 6.3). The visually brightest 'pure' G star, one without a companion, is η Bootis (Mufrid, G0 IV), near Arcturus, shining at magnitude 2.7. The most luminous known of the class is a 5th magnitude G4 Ia supergiant in western Cassiopeia known simply as HR 8752 (see Figure 6.4) in the Bright Star Catalog, or as HD 21776. As is common for ultraluminous stars, it varies irregularly, with a range of about half a magnitude, and is also called V509 Cas.

6.1 Spectra

As we pass from the cool K and M stars into the intermediate temperature range, occupied at the low end by the G stars, the lines of the neutral metals, such as $\lambda 4226$ of Ca I and the sodium D lines, weaken, although those particular ones are still quite strong and very noticeable (Figure 6.5). The spectra of singly ionized metals are quite strong. Fraunhofer H and K of Ca II attain their greatest strength as class K merges into

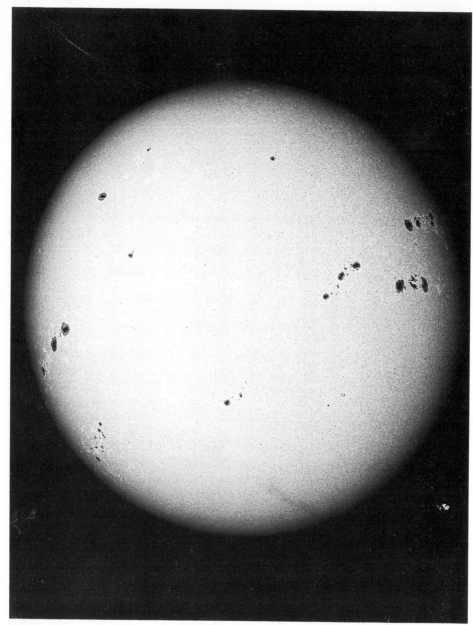

Figure 6.1. The Sun: a close view of a G dwarf. The bright surface is the photosphere, and the spots mark the centers of activity (Section 6.6). The whitish areas are bright zones in the overlying chromosphere (Section 6.4). Note that the center of the Sun is brighter than the edge, which is a result of an increase in temperature toward deeper layers (Section 6.8). Mt. Wilson and Las Companas Observatories, Carnegie Institution of Washington, photograph.

Figure 6.2. Capella (α Aurigae), whose spectrum is dominated by a G5 giant, and her three Kids. Epsilon (K4 II + B8 V), in the center, and Zeta (whose primary is F0 Ib), lower right of the three, are the famous eclipsing binaries that were discussed in Section 5.4. The third star of the trio, at lower left, is Eta, a B3 dwarf. Lowell Observatory photograph.

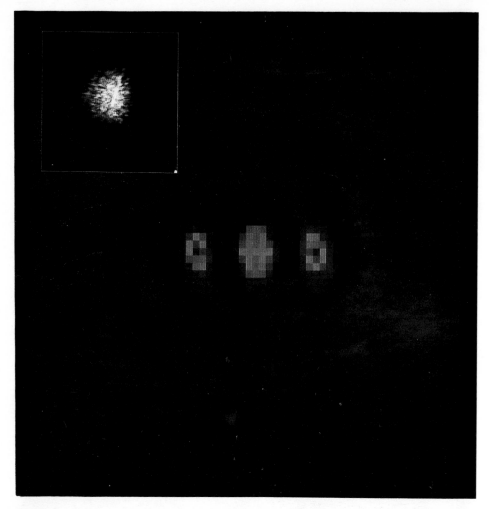

Figure 6.3. Capella resolved. This close G0 III G5 III double, whose components are only 0.043″ apart, is not separable by eye. It is resolved here by a technique called 'speckle interferometry' (Section 4.4), in which a large number of very short exposures are taken, freezing the star in time as it twinkles. In the inset at the upper left, we catch the many tiny individual images, each one a double, that constitute the large blurred object we would otherwise see. The central figure shows the final computer-processed image of Capella derived from such observations. The brighter G5 giant is at the center. The G0 companion appears twice, once on each side. One is real, and the other is a 'ghost image' that must be eliminated by other techniques. This method was used in the uncertain, and probably false, detection of the brown dwarf companion to the M dwarf VB 8 (Section 4.10), which illustrates the uncertainty and difficulty of the technique. National Optical Astronomy Observatories (Kitt Peak) and Georgia State University, courtesy of H. A. McAlister.

Figure 6.4. The HR diagram for classes G and K. We begin to see some Cepheid variables (Section 6.11) in a band (the shaded zone) in which stars become unstable, and which will continue on into type F. Note RV Tauri (Section 6.11), which is located in its average position. The Sun falls on the main sequence not far from its neighbor α Cen A. We also see both giant components of Capella. See Figures 3.8 and 12.7 for credits.

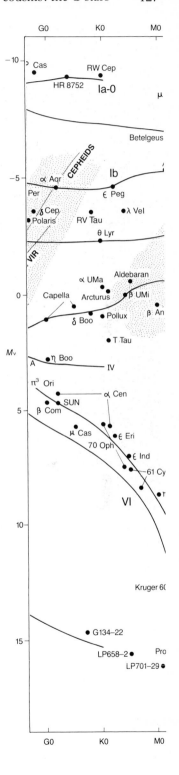

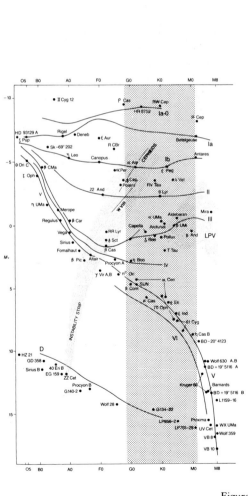

Figure 6.4.

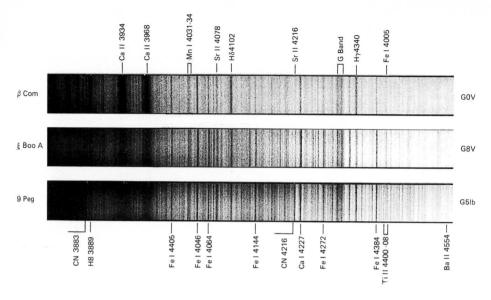

Figure 6.5. Spectral changes among the G stars. Variations through class G tend to be subtle: the differences between the G8 and G0 main sequence stars in the upper two strips are at first glance slight. But note how Hγ and Hδ strengthen as temperature increases, and how neutral calcium and iron diminish. The G band is also considerably weaker in β Comae Berenices than in ξ Bootis A. The lowest strip shows a G5 Ib supergiant spectrum, intermediate in temperature between those of the dwarfs. Look at the strength of Sr II, which is barely visible on the main sequence, and at the development of the CN break at λ4216 (which has Sr II λ4216 at its longward edge), both of which are luminosity indicators. From *An Atlas of Representative Stellar Spectra* by Y. Yamashita, K. Nariai, and Y. Norimoto, University of Tokyo Press, Tokyo; John Wiley & Sons, New York, 1978.

G, and dominate the violet spectra of stars like the Sun. The hydrogen lines become among the more prominent of features as higher temperatures elevate additional electrons into the second energy state from which the optical Balmer series arises: see Sections 2.11 and 3.7.

One of the better indicators of Draper type is the strength of Hδ relative to that of Fe I at λ4144. The two are about the same at G8, and the ratio increases very noticeably toward earlier subtypes. The positive luminosity effect seen for hydrogen in classes K and M all but disappears at G, which gives us good consistency in classification among the dwarfs, giants, and supergiants. Line ratios of various metals, such as chromium to iron, are useful as well.

Luminosity classification, however, becomes a bit tricky for the G stars as there are few clear indicators. The ratios of the strengths of various metallic spectra are used, particularly those of ionized strontium to neutral iron (as the ionized lines are enhanced at higher luminosity). The Ca II H and K lines also strengthen with decreasing absolute magnitude. An accurate luminosity can be deduced from the Wilson–Bappu effect (Section 5.3) if the star is bright enough to allow us to detect and measure the width of the emission core of the K line. The method, however, requires high-dispersion spectra and long exposures, and is not easy to apply.

Class G is about the last we see of molecular spectra with any significant strength. At these temperatures of 5000–6000 K only the hardier molecules survive. The G band, with its lines of strongly-bound CH, reaches maximum strength at G5, and then weakens toward type F as even these bonds break under the stress of more energetic atomic collisions. The CN break at $\lambda4215$ has weakened to invisibility in G dwarfs, but because of its positive luminosity effect, it is seen in late G giants and supergiants. The observer must be very careful in using it for luminosity classification, however, because of the strong variations that can be induced by carbon and nitrogen abundance anomalies.

6.2 The solar spectrum

In each of the two previous chapters we looked at stellar spectra taken with high dispersion, those of Betelgeuse and Mira for the M stars (Figures 4.3 and 4.8), and of δ Tauri and Arcturus for the K (Figures 5.2 and 5.4), in which we can see considerable structure. For the G stars we have our Sun, and we have so much light that we can disperse or stretch the spectrum to an astonishing degree (Figures 6.6, 6.7, and 6.8), allowing the finest detail to become visible. We can also examine the solar spectrum in a variety of temperature regions: the cooler chromosphere and sunspots, and the hotter outer corona (see below). In all, roughly 50 000 absorption lines and a wide variety of emission features have been detected. Consequently, the solar spectrum is a fundamental reference for other stars, and the chemical composition derived from it is a basic standard to which the rest of the Galaxy is compared.

Of the 92 natural elements, 68 have been found in the Sun (Figure 6.9). One, helium, was discovered there before it was isolated on Earth, hence its name from the Greek 'helios' for that body. Of the missing 24, ten are highly radioactive, and long ago disappeared completely from the solar system. Recall from Section 4.7, though, that technetium (Tc, no. 43) is known in S, C, and even in M-type stars, where it is freshly

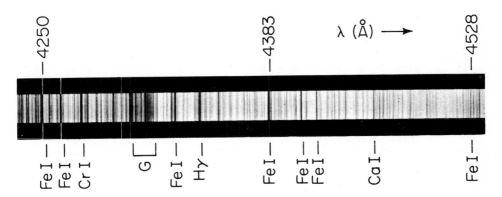

Figure 6.6. A moderate dispersion spectrogram of the Sun, between 4230 and 4540 Å, in the neighborhood of Hγ and the G band of CH. The Hγ hydrogen line, although prominent, is no stronger than many of the numerous iron lines. From the *Binary Stars*, by R. G. Aitken, McGraw-Hill, New York, 1935.

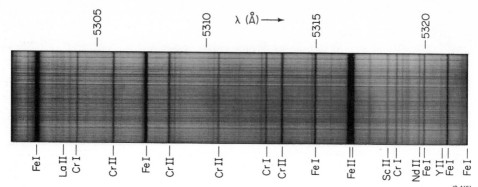

Figure 6.7. A high dispersion solar spectrogram, spanning 23 Å in the yellow, which shows enormously more detail than the previous spectrum. The slit of the spectrograph is held steady on the Sun, and the horizontal striations are caused by granulation (see Figure 6.11). The irregular wiggles in the absorption lines are the result of vertical motions that produce minute point-to-point Doppler shifts, from which velocities can be derived. Mt. Wilson and Las Campanas Observatories, Carnegie Institution of Washington, photograph, courtesy of E. C. Olson.

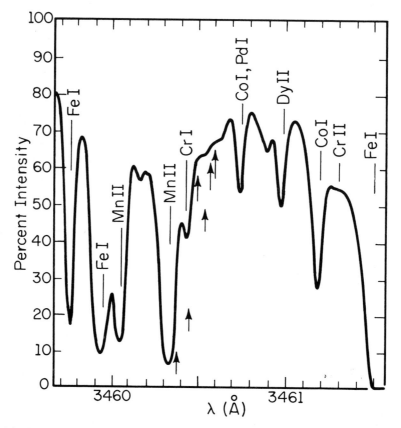

Figure 6.8. An extremely high dispersion scan of the solar spectrum only 2 Å wide, which shows the remarkable detail that can be attained. The difficulty of identifying lines of low-abundance elements is illustrated by the arrows, which indicate expected positions of lines of the rare element rhenium (no. 75), which has not yet been found. The blending of common lines is so severe that the rhenium absorptions, if present at all, are lost. Adapted from an article by J. W. Swensson in *Solar Physics*, vol. 18, D. Reidel Publishers, Dordrecht, Holland.

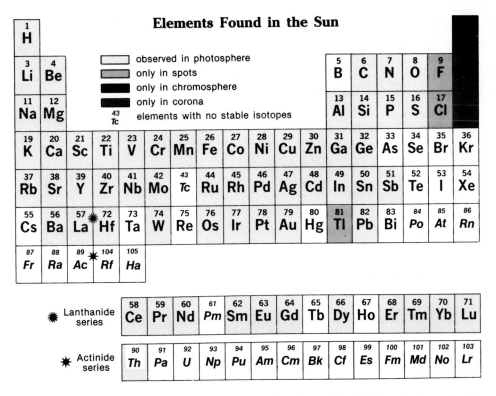

Figure 6.9. A periodic table of the elements, indicating those observed in the Sun, coded according to whether they are detected in the photosphere, or *only* in the chromosphere, corona, or sunspots. Those not observed are left unshaded and radioactive elements are set in italics. Atomic numbers are given above the chemical symbol. Diagram by the author.

manufactured in the interior and cycled to the surface. The remaining elements are certainly present in the solar gas; they are not detected primarily because their rarity results in very weak lines that are blended and lost within the stronger features of the more common atoms (Figure 6.8). The great solar luminosity also allows us to see numerous, often very weak, molecular lines. In addition to CH we observe spectra from, for example, OH, NH, O_2, and CO.

An immense effort has gone into the determination of relative solar abundance ratios. These are found from the relative strengths of the absorption lines, from our knowledge of how the lines are formed (see Chapter 2 and Section 4.3), and from a vast mass of laboratory data that gives us information on the probabilities of any given bound electron making any of the transitions that create the observed features. The results are seen in Figure 6.10, in which relative abundances are plotted against atomic number. We believe this distribution to be typical of the stars in our part of the Galaxy, and of the galactic disk in general.

There are a number of features to note in the graph: first, the overwhelming abundance of hydrogen, followed by an overall decline with increasing atomic number to very low values; next the great paucity of the light elements lithium and beryllium

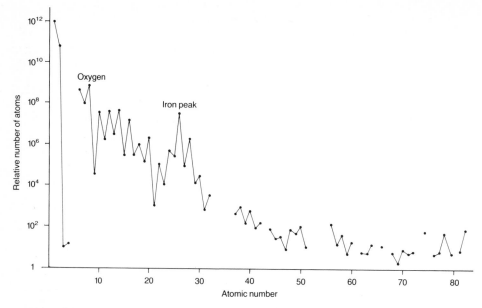

Figure 6.10. The relative abundances of the elements in the Sun, scaled to hydrogen at 10^{12}. The graph is the result of a compilation by J. E. Ross and L. H. Aller of over 100 technical articles. Lithium is reduced by a factor of 100 as compared to its abundance in the interstellar gas. Constructed from data in *Science*, vol. 191, published by the American Association for the Advancement of Science, New York.

(boron has yet even to be found); third, the alternating pattern of elements, where those with even atomic numbers always have abundances larger than their odd-numbered neighbors; and finally, a very distinct peak at iron (no. 26). These variations can be understood through a discipline known as *nuclear astrophysics*, in which we show how heavier elements are created in stars from lighter ones by thermonuclear fusion, beginning with hydrogen and helium as the initial building blocks. The process itself is called *nucleosynthesis*. The irregularities are caused by varying probabilities of formation of the different elements, and by the different strengths of the various nuclear bonds.

This is the mixture with which the Sun began its life. The heavier elements were all created in the cores of other stars and spewed into the Galaxy by winds and stellar explosions long before the Sun was born. The evolution of stars and of the Galaxy and the differences between Populations I and II (Section 1.8) can thus be studied in part through the variations among stellar atomic abundances. It is of the utmost importance then that we search for deviations from the solar chemical mix, so that we may test our theories, and to provide data for the development of new concepts. Some specific examples stand out. As pointed out in Section 4.7, the R, N, and S spectra that we encountered earlier are created when by-products of thermonuclear fusion, including carbon, are flushed to the stellar surfaces by various circulation processes. Their study then yields information on what happens in the interiors of giants. The metal-poor subdwarfs and the weak CN giants of the galactic halo (Section 5.5) reflect the composition of this oldest part of the Galaxy, which was created before formation of the galactic disk. Comparisons between these stars and younger ones then let us explore in detail how the Galaxy has aged. We will encounter further fascinating discoveries in these research areas in later chapters.

6.3 Dwarfs and the Sun

In Chapters 4 and 5, we began with luminous giants and supergiants and worked our way downward in luminosity toward the dwarfs. We now reverse direction. The main sequence is becoming much brighter at type G, and the giant branch is converging on it. The dwarfs of this class now begin to make a real impact on our naked-eye sky.

Here we find stars that are very similar to our Sun (Table 6.1); presumably all G dwarfs share at least some of the characteristics of the Sun. We are really just beginning to see some of the solar features mirrored elsewhere. Our star – like the others – is not just a simple blackbody radiating into space, but has a complex, active, layered structure: see Sections 2.11, 4.9, and 5.3. We divide the visible Sun (and the cooler stars in general) into three distinct zones. The lowest of these, the photosphere, produces the bulk of the radiation; it is the so-called solar 'surface', a layer of very opaque gas. The high opacity, which gives the Sun its sharp-edged appearance, is produced by a *negative* ion of hydrogen, H^-: a proton with *two* orbiting electrons. The second is easily stripped away by the outgoing solar radiation, and consequently absorbs and blocks it very efficiently, limiting the depth to which we can see.

A properly filtered view of the photosphere shows its granular, turbulent nature (Figure 6.11). We see tiny blobs of gas set within a cooler matrix that come and go and continually re-form themselves. Doppler shifts in their spectra shows the gases to be continuously rising and falling (Figure 6.7). The granulation represents the end

Table 6.1. *Other suns*

What would our Sun look like from a great distance? Here is a short list of named naked eye stars classified at or very near G2 V. 16 Cyg A and B are classified as G1.5 V and G2.5 V respectively. The closest match to the Sun is 9 Ceti. Note the large spread in absolute magnitude, which gives breadth to the main sequence.

	Apparent visual magnitude	Distance in parsec	Absolute visual magnitude
53 Aquarii A	6.57	18	5.35
α Centauri A	−0.01	1.33	4.37
9 Ceti	6.39	20	4.84
ϱ Coronae Borealis	5.41	25	3.42
16 Cygni A	5.96	26	3.92
16 Cygni B	6.20	26	4.16

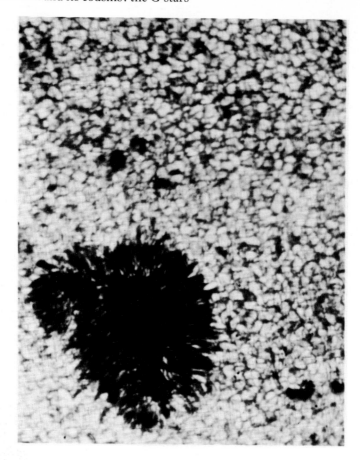

Figure 6.11. A very close view of a sunspot, a feature of the active Sun, and of the photosphere. The spot is a magnetically cooled zone of complex structure. The solar surface shows a delicate fine structure called *granulation*, which is related to convection in the lower solar layers. The small black areas are called 'pores', and may be nascent spots, or may quickly disappear. Photograph from Project Stratoscope, Princeton University, supported by ONR, NSF, and NASA.

product of a deep zone in which the matter keeps churning over by convection – the process by which the hot portion of a fluid will rise, and the cooler part will fall.

6.4 The solar chromosphere and the corona

On top of this bottom layer is the low-density, 12 000 kilometer thick chromospheric transition zone that we encountered among the M and K stars in Sections 4.9 and 5.3 (Figure 6.12). The temperature is about 1500 K cooler here than in the 5800 K photosphere, but begins to climb as the outer boundary is approached, shooting up to well over one million degrees in the great outer envelope, the corona (Figure 6.13). This awesomely lovely solar crown is marvelously complex, with polar and equatorial streamers, holes, loops, and whorls. The temperature can reach 10 million degrees; but because of its extremely low density it, in company with the chromosphere, does not

Figure 6.12. The chromosphere, shown in a highly magnified, high resolution, *spectroheliogram* of the Sun. In this technique we photograph the Sun in the center of one of the strong absorption lines, usually calcium K or Hα. The gas is so opaque at these wavelengths that our vision cannot penetrate to the photosphere, and we record only the cooler chromospheric gases. This picture, taken in the light of Hα, shows a field of *spicules*, jets of gas that spout out of the chromosphere into the lower corona. This layer of the Sun is in a constant state of change, as individual spicules last for only a few minutes. From the National Optical Astronomy Observatories.

Figure 6.13. The solar corona seen during a total solar eclipse. Note the polar streamers that outline the Sun's general magnetic field, and the large coronal loops. From the Naval Research Laboratory.

behave as a blackbody, and is optically so faint that it can be seen by eye only during a total solar eclipse. The temperature in these two layers refers only to the immense speed and energy of the individual atoms, which are so agitated that very highly ionized states can be produced: for example, iron with 15 of its electrons stripped away. The energetic corona is the source of the solar wind, the flow of high speed atoms that constantly blows past the Earth and other planets.

The corona is intimately connected with the Sun's magnetic field. Such fields are produced when electrified particles are set in motion. The solar magnetism is generated by the Sun's slow (monthly) rotation coupled with the circulating ionized gases in its deep convection zone that makes it behave as a vast dynamo. The magnetic lines of force drift outward along with the flow of solar energy, and can break through the surface in huge loops that seem to be responsible for containing the coronal gases. The source of the corona's heat is still being contended. There are two competing, and perhaps complementary, theories. The older says that acoustic (sound) waves are generated by the same convective cells that cause the granulation. These propagate outward, and as a result of the ever-decreasing density, are converted to shock waves much like the sonic booms produced by a supersonic jet plane. They then stir up, or heat, the corona. The alternative notion is that this outer halo is electrically energized by magnetic-field waves generated from below. Electrical pulses produced by the myriad explosive chromospheric flares (see Section 6.6) may have an effect as well. All mechanisms may in fact play a role: the chromosphere might be heated by the acoustic process and the corona by the magnetic, but the question is still open.

As optically faint as the Sun's chromosphere and corona are, they are still easily studied since the Sun is so close and angularly large that we can spatially separate all the layers. But when we examine other stars, the light from all the zones blends together into a tiny point, and the two weakly radiating outer ones become very hard to detect against the bright stellar photospheres. Nevertheless, they are found. We have already seen (Section 5.3) that chromospheres manifest themselves through central emissions in the Ca II H and K line profiles. Coronae have their own characteristics that make them viewable from afar. Their high temperatures cause X-rays to be emitted in copious amounts, which are detected by X-ray telescopes mounted aboard Earth-orbiting satellites.

6.5 Stellar chromospheres and coronae

The examination of chromospheric and coronal phenomena in other stars allows us to test the dynamo theory of the production of solar and stellar magnetism. Recall from Section 5.3 that a stellar chromosphere produces emission line cores at the center of Ca II H and K. We observe that the energy radiated in these lines increases as we proceed down the main sequence, implying that the prominence of the chromosphere relative to the photosphere does as well.

Theoretically, we expect that the magnetic fields should strengthen with both an increase in stellar rotation speed and with an increase in the depth of the convection zone. The Sun is a relatively slow rotator among stars. Simple observation of the motion of dark spots across its face shows that it makes a full turn in about 25 days, with an equatorial speed of 2 km/s. We can determine the rotation speeds of other late-type stars by the fact that chromospheric emission from stellar active (see Section 6.6 below) regions scattered over their surfaces waxes and wanes as rotation moves the emitting zones in and out of sight. Even among the so-called 'field' G stars (like our Sun, those not allied with clusters) the Sun still turns slowly, as 5 km/s is more typical. Within rather wide margins, the cooler field dwarfs spin the more slowly. In early K the speed averages about 2 km/s, dropping closer to one in late K.

On this basis alone we might expect the relative importance of the chromospheres, as evidenced by their K-line emissions, to diminish rather than strengthen as we progress downward along the main sequence. However, stellar structure theory shows that the convection zones deepen as stellar mass decreases toward the later types. This effect overrides that of the slower spins, and causes the chromospheres and coronae to strengthen relative to that of their respective photospheres. Although the theory is beset with serious and confounding difficulties it seems correct in broad outline.

Stellar coronae, as studied through their X-ray radiation, are also consistently present with increasing strength relative to total stellar luminosity from late A through M. Beyond M5, however, the X-ray incidence drops off for reasons not yet clear. There also seems to be a transitional region toward the upper right of the HR diagram into the realm of the red giants and supergiants where the coronae disappear, related perhaps to much slower rotation speeds, to the diminishment of the convective layer, or to the onset of massive stellar winds (Section 4.6).

The ages of stars play an important role in the incidence of stellar chromospheres and other phenomena. Stars of different ages are easily identifiable through their memberships in clusters. As we have already noted (and will see in greater detail later), the lifetime of a star shortens dramatically as we proceed up the main sequence in the direction of higher mass, since the available fuel is consumed with much greater speed. The most newly formed clusters, those whose stars are very young, can then easily be found by the presence of O and B stars, and the older clusters by their lack.

The older stars exhibit less chromospheric activity and lower rotation speeds, which presumably are coupled together. The magnetic fields that are generated by rotation are carried outward by the ionized gas of the stellar wind. Since the outflowing gas cannot continue to rotate with the same period as the star, the connecting lines exert a small but inexorable force that gradually slows the star down, causing the chromosphere to diminish. We will examine stellar rotation and the phenomenon of magnetic braking in more detail among the F stars.

6.6 Solar activity

Superimposed upon this layered solar structure is a set of active, ephemeral features that can approach the truly bizarre. We have known since Galileo's time that the Sun is afflicted with dark spots (Figures 6.1 and 6.11), ranging from mere pinpoints to areas large enough to be seen with the naked eye during a hazy sunrise or with proper filtration. (Warning: NEVER look directly at the Sun unless you know exactly how to do it, as you can be blinded.) These spots are about 1500 K cooler than the surrounding photosphere, and have individual lifetimes of days or weeks. Their numbers increase then decrease over an interval called the solar cycle, which averages about 11 years in length. At the cycle's peak, the Sun is quite covered with them, while at minimum it can be quite difficult to find any at all. They are actually but one manifestation of a general 22 year magnetic cycle, or period of solar magnetic variability. The spots themselves are regions of intense magnetism that is thousands of times stronger than the Earth's field, and which appears to inhibit the flow of radiation from below, rendering the zones cool and relatively dark.

The spots generally come in pairs, and one or more of these duos are grouped into a larger magnetic area called a *center of activity*. The magnetic field that causes the phenomenon loops out of the photosphere at one spot and back in at the other, so that each of a pair has a different polarity, that is one is positive (a 'north' pole) the other negative (or 'south'). Powerful evidence shows that the spot activity is controlled by the Sun's general magnetic field. During one sunspot cycle of 11 years this field may be aligned with the rotation direction, that is, magnetic and rotational north and south coincide. In that case, the leading spot of a pair on the northern hemisphere (the one that is ahead in the direction of rotation) has a positive magnetic polarity, and the follower is negative. In the southern hemisphere, these directions are reversed. During the next 11-year period, the general field is switched north-for-south, and so are the directions of the spot polarities.

A spot cycle begins from minimum with a few found at mid-solar latitudes. As it progresses, we see more and more of them, and their average position moves closer to the equator. Finally, after the cycle has played out, we will see only a few left at very low latitudes, and the new magnetically reversed period takes over.

The name 'center of activity' is truly descriptive. Above the spots, we may see *solar flares* (Figure 6.14), intense explosive regions in the chromosphere caused by a sudden release of magnetic energy. They seem to be allied with the explosive brightenings observed among the M dwarfs (Section 4.9). These events are also associated with great bursts of radiation seen at radio wavelengths.

The corona displays other activity-related features. There we see dark sheets of gas projected against the photosphere; when these *filaments* are outlined against the sky at the solar limb, they appear as the bright *prominences* (Figure 6.15), which can put on outstanding displays. Time-lapse photography shows some prominences that rain matter down onto the chromosphere, whereas others are explosively violent, hurling streams of matter into space.

A current view of the solar cycle is that it is caused by the Sun's magnetic field being sheared and twisted by differential solar rotation. Like most gaseous bodies, the Sun does not spin uniformly. The basic rotation period of 25 days refers to the equator. The interval increases to near 30 close to the poles, so that the equatorial gases move past those at higher latitudes. Since the magnetic field is locked into the electrically charged, ionized matter of the photosphere, the rotation distorts the field, slowly wrapping it around the Sun. Forces that are still only vaguely understood can cause specific areas of the field to float upward and break through the surface, at which point a center of activity is created. These magnetic loops are related to those that contain coronal gases, and consequently, the appearance of the corona is also correlated with the activity cycle. As the twisting proceeds we see more and more spots, until the field is so tightly wound that it begins to break apart. The number of spots decreases, the now-disordered field reorganizes itself with a reversed direction, and after an interval of 11 years the whole process begins again with a new minimum. A newer idea holds that activity is related to huge convection cells that drive the spot zones closer to the equator, and that two twenty-two year cycles proceed concurrently, one 11 years behind the other, with the spots developing only during the second half. An immense amount of labor remains to be done before we will understand this enormously complex phenomenon.

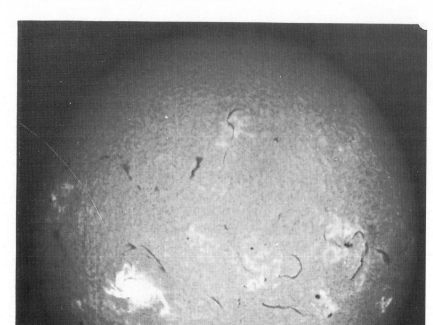

Figure 6.14. This Hα spectroheliogram shows the large-scale mottling of the chromosphere, as well as more features of the active Sun. At the lower left is a large flare, a chromospheric zone associated with an active region, in which stored magnetic energy is explosively released, with the copious production of X-rays. The dark, snakelike lines are filaments, sheets of gas that hang in the corona. From the McMath–Hulbert Observatory of the University of Michigan.

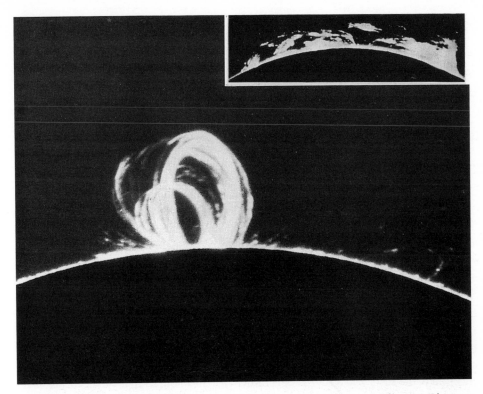

Figure 6.15. Two examples of the great variety of prominences, which appear as filaments when seen projected against the Sun (see Figure 6.14). The large photo shows a loop prominence, which is formed by a strong magnetic field that shapes the hot gas into this distinctive configuration. The inset at upper right shows quiescent prominences that develop above sunspot groups. Time-lapse photography shows that the gas is streaming downward onto spots at the center. Some other prominences will show little motion, while others erupt explosively. Photos are respectively from National Optical Astronomy Observatories, and the University of Colorado High Altitude Observatory.

6.7 Stellar activity

Similar activity cycles associated with other solar-type dwarfs can be detected. The level of activity in a star strongly affects the chromosphere, and consequently the strengths of the H and K line emission cores. In addition to the short-term modulation of the chromosphere caused by rotation, and the movement of active zones into and out of our sight, we have strong evidence for long-period solar-type cycles among late-type stars that range from a few years upward. Although no great area of the Sun is very covered by spots at any one time, they will also produce a very slight variation in total luminosity that would in fact be too small to be detected were the Sun at a great distance. However, a few of the stars that show chromospheric activity cycles do produce such fluctuations; they seem to be quite covered with starspots. Recall also the flare activity observed for type dMe (Section 4.9), which probably represents an extreme variant of solar behavior.

6.8 The Sun as a star

Of inestimable importance is our ability to measure precisely for the Sun a variety of fundamental astrophysical quantities (Table 6.2). Our star is so close and sharply defined that we can determine its radius to within a fraction of a percent; it is a difficult feat to get any measurements at all of this kind for any other star. Since it is so bright, we can obtain the Sun's luminosity to a similar precision. The combination of size and brightness allows us to find the effective temperature to within a few degrees (Section 4.3), an accuracy impossible to approach for our more distant stellar neighbors. From the period and radius of the Earth's orbit and Kepler's third law (Section 1.6), we can deduce the Sun's mass to within one part in ten thousand. We can make direct measurements and take samples of the solar wind, whereas similar phenomena can only be inferred from the spectra of other stars. Surface fluctuations and gas motions can be monitored to within a fraction of a kilometer per second, and subtle moment-to-moment and point-to-point variations in the magnetic field strengths that are so important to studies of solar activity can be followed. Magnetism in other stars cannot even be detected unless the total field strengths are hundreds of times larger than they are on the Sun.

We can even penetrate into the Sun and explore the layering of the photosphere, since the solar gases are partially transparent. This body looks like a flat disk in our sky, but in reality, it is of course spherical. The limbs, or edges, are over half a million or so kilometers farther from us than is the center. At the periphery, our line of sight encounters the surface at a sharp angle rather than head on, so that we see to shallower and consequently cooler layers than when we look at the center. Because a hot gas radiates more light than a cool one (Section 1.10), the edge of the sun is then very noticeably darker than the middle. The observations of this limb-darkening, which is

Table 6.2. *Some solar data*

	Physical units	Relative to Earth
Diameter	1.392×10^6 km	109.2
Mass	1.989×10^{33} gm	3.328×10^5
Average density	1.409 g/cm^3	0.2553
Surface gravity	2.740×10^4 cm/s^2	27.94
Escape velocity	617.7 km/s	55.20
Polar magnetic field	—	5[a]
Luminosity	3.83×10^{26} watts[b]	
Apparent visual magnitude	-26.74	
Absolute visual magnitude	$+4.83$	
Color index (B−V)	0.65	
Color index (U−B)	0.13	
Bolometric correction	-0.08 magnitudes	
Effective temperature	5770 K	

[a] at solar sunspot minimum – value only approximate
[b] a watt is 10^7 ergs per second

very noticeable in Figure 6.1, can be used to reconstruct the progress of temperature with depth over a distance of several thousand kilometers. Thus the Sun provides an observational reference against which to compare our theories of the construction of stellar atmospheres.

6.9 The solar interior

Our 'vision' also takes us past the atmosphere directly into the solar interior. All stars, except for the burned-out white dwarfs and neutron varieties, are powered by the processes of thermonuclear fusion. We first explored this subject in Section 1.16. At the temperatures found in main sequence stellar cores, four atoms of hydrogen are fused into one of helium, or in giants three of helium into one of carbon, with a resulting release of energy. The specific mechanism in lower main sequence stars (about F2 and later) involves the successive collision and addition of protons into light helium, followed by the merger of two of these into normal helium-4. This *proton–proton* (or *p–p*) reaction is outlined and discussed further in Display 6.1. The high temperature (15 million K) of the core is needed in order to give the protons enough velocity so that they can overcome their mutual repulsion to approach one another closely enough to allow them to stick together via the strong force (Section 2.1).

We know of the process by inference, since with light or radio waves we can look only into a star's surface layers, far short of its core. Calculations of theoretical stellar models show that fusion reaction rates calculated for stars of different masses produce the observed stellar luminosities, and therefore our theories are likely to be correct.

However, the internal nuclear fusion also generates sub-atomic particles called *neutrinos* (see Display 6.1). Practically everything is transparent to them – an average one can penetrate a light-year of lead – and thus they escape the solar core immediately.

Display 6.1. *Solar power*

The energy of the Sun, and of all main sequence stars later than about F0–F2, is generated mostly by the proton–proton reaction:

$$^1H + {}^1H \rightarrow {}^2H + \gamma + \nu$$
$$^2H + {}^1H \rightarrow {}^3He + \gamma$$
$$^3He + {}^3He \rightarrow {}^4He + 2({}^1H) + \gamma.$$

The numerical superscripts refer to atomic mass (or isotope). The helium atom is 0.7% lighter than the four H atoms are separately. The lost mass is converted into energy by the production of gamma rays (γ) and neutrinos (ν) via the famed equation $E = Mc^2$. Because the speed of light, c, is so great, a very large amount of energy can be created from a small amount of mass. (One gram of matter converted completely into energy could light a *billion* standard light bulbs for an hour).

The top reaction actually produces a positron, a positive electron that is a form of antimatter. This particle quickly encounters a normal electron, and the two annihilate one another to generate the energy. A secondary chain, not shown, produces beryllium, and creates the higher energy neutrinos that are detected by the neutrino telescope discussed in Section 6.9.

But because neutrinos are produced in such vast quantities, and the Sun is so close, we have (with difficulty) been able to build an effective detector on Earth. This neutrino telescope, which can literally see the reactions taking place in the solar center, consists of 400 000 liters of a chlorine-based cleaning compound buried in a deep mine in South Dakota. On rare, but statistically predictable occasions, a chlorine atom can capture a neutrino that is produced by a side-chain to the p–p reaction (see Display 6.1) and undergo a nuclear transformation into radioactive argon. The atoms of the latter can be detected and numbered. By knowing the probability of capture, which of course is terribly low, and the relation between the number of side-chain neutrinos and those produced by the main reaction, we can count their total flow.

What we find is somewhat disconcerting: fewer neutrinos seem to be coming out of the Sun than we expected. We do not know if the problem is with our solar models, with the experiment itself, or whether changes have taken place in the solar core that are not yet reflected at the surface. More sophisticated experiments that can count the number of lower energy neutrinos directly may resolve the matter. One exciting possibility involves our understanding of the nature of neutrinos themselves. There are different varieties of these tiny particles, and they are generally believed to have no mass. However, if they do have minuscule masses, they can conceivably switch from an observable type to one that is invisible to the present equipment, thus resolving the discrepancy. The question of whether or not neutrinos can actually have mass is a profound one. There are so many flying about the Universe that they would constitute a major portion of its mass, with deep cosmological implications. In any case, as usual, it is our own local G star that is leading the way toward our ultimate comprehension of stellar physics.

6.10 The aging of the Sun and stars

At the end of Section 6.5 we looked briefly at how stellar rotation and the incidence of chromospheres dimish with age. Now let us examine the subject of aging again by looking at nuclear reactions and, perhaps curiously, lithium abundances. This light element, number three in the periodic table, is anomalously deficient in the Sun: its abundance is only one percent of that of some earlier type stars and of the interstellar medium out of which the Sun formed some five billion years ago. It is quite certain that solar lithium has been depleted over time. Observations of strength of the Li $\lambda 6708$ resonance doublet (Sections 2.9 and 2.11) in a broad sample of stars show that it weakens as we proceed down the main sequence from class F to very early K, where it disappears. There is also a loose correlation with age for a given spectral type (Figure 6.16). The line is stronger in stars that belong to younger clusters and in those with more rapid rotation and with greater chromospheric activity, which have not yet been magnetically braked (Section 6.5).

Lithium is easily destroyed by thermonuclear processes. We think that it is cycled downward by convection into deep, hot layers where each atom is converted by capture of a proton into two helium nuclei (the reaction is unimportant in energy production). Therefore later type stars with the deeper convection zones will have less of it, and of course the amount will also go down with time.

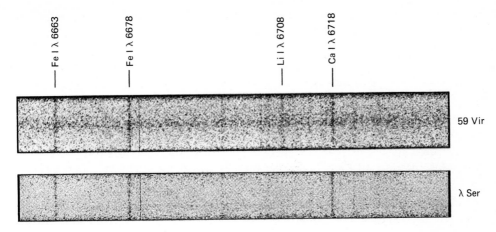

Figure 6.16. Lithium in stellar spectra. These two G0 V stars have identical spectra except that 59 Virginis has a fairly prominent lithium line (Li I λ6708) and λ Serpentis does not. The latter, like our Sun, is presumably older than the former, and has had its lithium destroyed by thermonuclear processes. The lithium line in 59 Vir will gradually disappear with time. Once the correlation with age is established, we can use the line to identify young field stars. Lick Observatory spectrograms from an article in the *Astrophysical Journal* by G. H. Herbig.

The problem is that according to current theory the convection zones of stars like and hotter than the Sun, G2 and earlier, are not deep enough to bring the element into regions with sufficiently high temperature so that we have to invoke other, somewhat arbitrary, circulation mechanisms. The subject of lithium will arise again in the next chapter, and is an important clue to star formation, as we will see in Chapter 12.

6.11 Subgiants, giants, and supergiants

We now explore upward in luminosity, and first encounter a relatively numerous set of subgiants about two magnitudes brighter than the Sun. These stars between the giants and the dwarfs are far less common within class K, and are absent altogether at class M. They embrace an interesting type of eclipsing binary (Section 5.4) called *RS Canum Venaticorum* (RS CVn) stars, which are also variable outside of eclipse. Both are usually G (or sometimes F or K) subgiants or bright dwarfs. As a result of tidal forces, the components rotate in synchronism with their revolution, and therefore they spin quite rapidly. According to the dynamo theory (Section 6.5), these stars should exhibit extremes of solar-like activity, and indeed they do. We think that the variation during the non-eclipse portion of their revolution is caused by extensive star-spot areas that are unevenly distributed on one or both stars due to the proximity of the other, and therefore rotate in and out of view. The pairs also send us strong X-ray emission from well-developed coronae, and even display occasional powerful flares analogous to those seen on the Sun and on the surfaces of the UV Ceti stars of type M (Section 4.9).

Another two magnitudes upwards on the HR diagram brings us to the G giants. The difference between dwarf and giant is now rapidly diminishing: a G2 III star is typically only six times larger than the Sun. Of course the supergiants can still be of stupendous size: HR 8752 (G4 Ia–0), at the top of our HR diagram, is over 500 times the solar diameter, and would more than encompass the orbit of the Earth.

As we move upward into the realm of the supergiants, we have our first meeting with the famed Cepheid variables, which have assumed immense importance in astronomy over the past many decades. They are pulsating stars with almost perfectly regular periods. Most of them, including the prototype δ Cephei, are ordinarily of type F, so we will discuss them in detail in the next chapter. But the most luminous ones, with M_v brighter than −4 or so, are G stars, and it is appropriate at least to mention them here. At their greatest extreme, the set even extends slightly into early K.

More peculiar to class G are the rare *RV Tauri* variable stars. These Ib–II supergiants have light curves with large amplitudes of one or more magnitudes, which also display alternating deep and shallow minima. They have relatively long, fairly regular oscillation periods in the neighborhood of 100 days with spectra that range over types late F to early K. The stars bear some resemblance to the red Mira variables in that moving shock waves form in their pulsating atmospheres, and they seem to be losing mass at fairly high rates through strong stellar winds. The phenomenon may be related to a lowered metal content: they tend toward Population II, and some are seen in the oldest members of that set, the globular clusters. They may be the immediate precursors to some of the planetary nebulae (Chapter 11).

Equally rare and even more metal poor are the *yellow semi-regular variables*, sometimes called 'SRd' stars, which exhibit long-period but more erratic light curves of generally lower amplitude. There is still a great deal to learn about the few dozen known members of these classes.

We now pass our solar benchmark, and move on to the more luminous stars that occupy the upper main sequence. Before we do, however, let us pause to admire the Sun once again. As we develop more and more powerful instrumentation and space-based observatories, the amount of data gathered on individual stars will very rapidly increase. But in spite of how well we will ever understand the workings of, say, Arcturus or Vega, our own Star will always lead the pack, and provide the vanguard for research into the rest, as we attempt to find the analogues of solar behavior elsewhere.

7 Class F: stars in transition

The winter skies of the northern hemisphere are enlivened by the flash and sparkle of the brightest of all stars, Sirius. Atmospheric refraction combined with low altitude makes it jump and constantly change color, so that it is noticed by nearly everyone. The Dog Star, named for its location in Canis Major, was so important to early astronomers – its first morning appearance predicted the flooding of the Nile – that it has two stars that proclaim its coming. One of these is Murzim (β Canis Majoris B1 II–III), the Announcer, about 6° to the west. The other is first magnitude Procyon, whose name also shows obeisance: pro (before) kyon (dog), in Greek.

It is fitting, then, that yellow-white Procyon should also precede and announce Sirius physically as well: the former is our representative F star (F5 IV), and the latter is that great archetype of class A. Spectral type F is one of transition, and in several ways the entire class is a prelude to the A stars: here the molecular lines finally disappear; on the main sequence, rotation speeds suddenly increase, different and more efficient nuclear reactions begin to generate the stars' luminosities, and we begin to see some of the strange abundance ratios for which type A is famous. Yet class F is not without its own singular distinction: it is the great bastion of the Cepheid variables, the pulsating stars that are critical to our comprehension of stellar evolution and intergalactic distances (Figure 7.1).

For those who live below 35° north latitude, another famous F star makes an appearance, riding its daily path about 20° south of Sirius, and nearly as bright: Canopus, an F0 Ib supergiant. And not too far away on the HR diagram, at F8 Ib, is perhaps the most famous of all stars, Polaris, the second magnitude beacon that marks the North Celestial Pole round which all stars appear to move. The most luminous known is probably ϱ Cassiopeiae (Figure 7.2), located about 3° southwest of β. This strange star can vary by over two magnitudes. At its fourth magnitude maximum, we see it as type F8 Ia with an absolute magnitude of nearly -10, but its downward excursions on the HR diagram have taken it into the mid-K range.

Main squence F stars abound (Figure 7.3). With the increasing luminosity of the dwarfs, a large number – about 500 – are visible to the unaided eye. Only three, however, are third magnitude or brighter: α Hydri at magnitude 2.9, and α Circini and π^3 Orionis, both at 3.2. The last is readily seen as the brightest of the string of stars in western Orion that represents the Hunter's upraised left arm and cloak. Gamma

Figure 7.1. The Small Magellanic Cloud, one of our two nearby companion galaxies. Its Cepheid variables, mostly of class F, helped to define the distance scale of the Universe. The great globular cluster 47 Tucanae is at the right-hand edge. From the ESO/SRC Southern Sky Survey, Royal Observatory, Edinburgh. Original negative from the UK Schmidt Telescope Unit.

Figure 7.2. Western Cassiopeia, with two F stars: β(Fl IV), the lower of the two surrounded by diffraction halos, and ϱ(F8 Ia), marked with the arrow. The latter is one of the most luminous stars known and is an odd variable that has been known to dim into type K. The bright star at the center is γ, a variable B0e IV star. Lowell Observatory Photograph.

Figure 7.2.

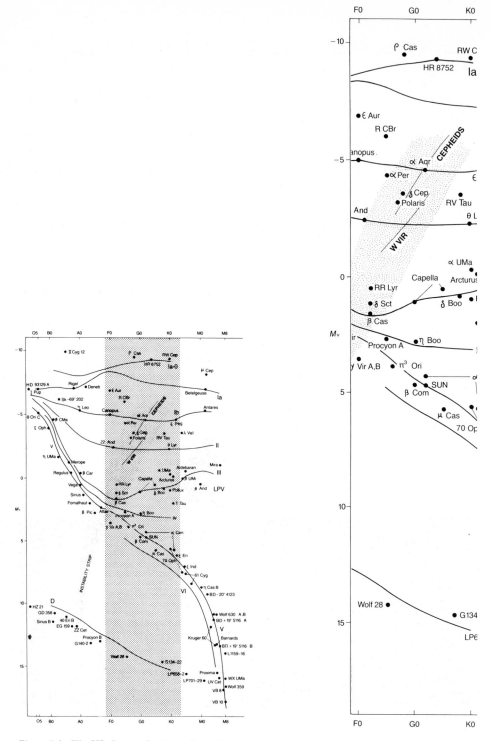

Figure 7.3. The HR diagram for classes F and G. The weird eclipsing binary ε Aur and the brilliant Canopus now make their appearance. The shaded area is the great instability strip where we find the Cepheids and the W Virginis pulsators. The strip will continue downward into class B. See Figures 3.8 and 12.7 for credits.

Virginis also appears to the eye at magnitude 2.9, but the telescope shows it to be a striking pair of F0 dwarfs a few seconds of arc apart, each of which singly is at magnitude 3.7.

7.1 Spectra

As we move into Draper F (Figure 7.4), we can unquestionably see the spectra becoming simpler as the numerous lines of the singly ionized metals slowly disappear, to be replaced by their doubly ionized versions whose absorptions are in the ultraviolet and are largely absent from the optical part of the spectrum. The hydrogen lines take on added strength, and except for the slowly fading H and K lines, clearly dominate. The G band of the CH molecule finally gives in to the ascending temperature, and by F3 is essentially gone.

The spectral type is best assessed by comparing $\lambda4227$ of Ca I, which is very rapidly fading with earlier subclass, with Hδ. The strength of what remains of the G band between F8 and F5 can also be useful. The luminosity class is found by comparing certain ionized and neutral metals, Fe II to Fe I, Ti II to Ca I or Mg I, Y II or Sr II to Fe I, where again the ionized states become enhanced relative to neutrals among the giants and supergiants as a result of low density (Section 3.7).

At higher dispersions (Figure 7.5 and 7.6) we can see some of the subleties of F star spectra. Look at the iron lines in the three spectra presented and the unusual elements noted in those of Procyon and α Persei. Here we also see the details of line shapes and the blending of closely-spaced lines in the spectrum of α Per caused by turbulent motions in that supergiant's atmosphere.

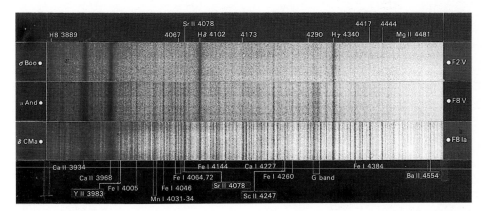

Figure 7.4. A selection of main sequence and supergiant spectra. Note the disappearance of the G band and the strengthening of the H lines toward early F in the dwarfs, and the loss of the G band and enhancement of ionized metal lines in the late F supergiant. Marked lines without identification are Fe II. From *An Atlas of Representative Stellar Spectra*, by Y. Yamashita, K. Nariai, and Y. Norimoto, University of Tokyo Press, Tokyo; John Wiley & Sons, New York, 1978.

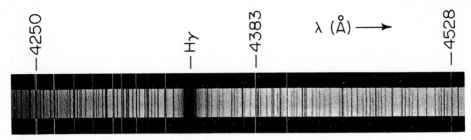

Figure 7.5. A moderately dispersed spectrum of the sky's second brightest star, Canopus (F0 Ib), in the neighborhood of Hγ. Note the iron lines by their alignment with the spectrograph comparison arc spectra displayed on either side of the stellar spectrum. Most of the other lines are produced by common metals such as titanium and chromium. From *The Binary Stars*, by R. G. Aitken, McGraw-Hill, New York, 1935.

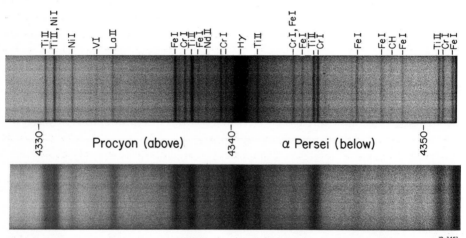

Figure 7.6. High dispersion spectra at Hγ of the famed F5 subgiant Procyon, contrasted with that of an F5 Ib supergiant, α Persei. Both of these superb spectra were taken with the coudé spectrograph of the 100-inch telescope at Mt. Wilson. The lines are again caused generally by common metals. However, note the singular appearance of vanadium, and especially that of the rare earth neodymium. A line of CH is also faintly visible in Procyon's spectrum. Compare these with the low dispersion F star classification spectra. Here, we isolate a tiny segment about Hγ, which reveals many more lines. The much higher dispersion also allows us to see that the supergiant's lines are broadened, an effect quite unnoticeable at low dispersion. The broadening is caused by Doppler shifts ascribed to poorly-understood mass motions usually referred to as 'turbulence'. Mt. Wilson and Las Campanas Observatories, Carnegie Institution of Washington, photographs, courtesy of R. and R. Griffin.

7.2 Dwarfs in transition: nuclear reactions

Two profound changes take place in class F, the first deep within the interior. The luminosity of a star is exquisitely sensitive to its mass (Sections 1.17 and 4.8). As the latter climbs, so does the central temperature, which causes the thermonuclear reactions to proceed at a greatly accelerated rate. From measurements of masses derived from binary star observations (Section 1.6), we find that the energy output of a main sequence star is very approximately proportional to the 3.5 power of its mass. That is, if you could double the mass, luminosity would jump by about a factor of ten.

In all main sequence stars, energy is produced by the fusion of four atoms of hydrogen into one of helium. (Giants and supergiants have other power sources, such as the already-mentioned fusion of helium into carbon; we will explore this subject more later.) In stars like the Sun, this process goes mostly by the *proton–proton* (*p–p*) *reaction* (Section 6.9 and Display 6.1), which involves direct H-atom collisions. But as we proceed to spectral types somewhat earlier than solar, the sensitivity of this nuclear reaction mechanism to temperature gradually drops well below that needed to maintain the observed mass–luminosity relation. The great brightnesses of the early-type stars of classes O, B, and A are generated by a different process, the carbon (or, sometimes the *carbon–nitrogen*, *CN*, or *CNO*) cycle, in which carbon acts as a nuclear catalyst (Display 7.1). The process requires high temperatures to make it run, so that it is unimportant in types M and K, and even in the Sun produces only about 5% of the energy. But its sensitivity to stellar heat is so great that it totally dominates in the early classes, satisfying the observed mass–luminosity relation. The competition between the p–p and CN cycles, of course, depends upon the carbon abundance. For a roughly solar mixture of atoms the crossover in the main sequence stellar power source occurs at about F0–F2, at a mass roughly 1.6 times that of the Sun.

Display 7.1. *The carbon cycle*

At about F0, the carbon cycle (sometimes called the CNO cycle) becomes more important than the proton–proton cycle (Display 6.1), and completely dominates in early-type stars:

$$^{12}C + {}^{1}H \rightarrow {}^{13}N + \gamma$$
$$^{13}N \rightarrow {}^{13}C + \gamma + \nu$$
$$^{13}C + {}^{1}H \rightarrow {}^{14}N + \gamma$$
$$^{14}N + {}^{1}H \rightarrow {}^{15}O + \gamma$$
$$^{15}O \rightarrow {}^{15}N + \gamma + \nu$$
$$^{15}N + {}^{1}H \rightarrow {}^{12}C + {}^{4}He + \gamma.$$

The carbon allows four H atoms to combine into one of He. ^{13}N and ^{15}O are radioactive isotopes that quickly decay into the next lower atom, and produce their γ-ray energy via the intermediary of a positron. Each reaction has many side chains. Neutrinos (Section 6.9) are denoted by ν.

7.3 Stellar rotation

Another great change involves the rotation speeds of main sequence field stars (those not in clusters). As we saw before (Section 6.5), G types typically spin at five or so kilometers per second, and K and M stars even more slowly. But rather suddenly at class F the typical speed begins to climb: 30 km/s at F5, 100 km/s at F0, up over 200 km/s within type B. An F0 star would present an interesting sight, completing a rotation in just under one Earth day.

The long rotation periods of the late-type stars are found from periodic variations in chromospheric emissions. At somewhat higher spin rates we can also begin to use the Doppler effect, which causes a widening of the absorption lines (Figure 7.7). As the star spins, one side approaches us, which causes the line to be shifted somewhat to shorter wavelengths relative to the line produced at the star's center. The other side recedes, causing a Doppler shift to longer wavelengths. The net effect is a smeared or broadened absorption caused by the gradually differing radial velocities and the resulting shifts in wavelength from the various parts of the star. Since all the lines are made wider in the same characteristic way, there is little confusion with luminosity or pressure effects. The positive and negative wavelength shifts, entered into the Doppler formula (Section 2.12) then yield the spin speed, and that coupled with an estimate of stellar diameter produces the rotation period.

The most difficult aspect of the rotation problem is with the orientation of the star's axis. If it is pointed directly at the observer, there can be no Doppler broadening no matter what the speed; we measure the full spin rate only if the axis is perpendicular to our line of sight. Ordinarily, what we derive is the equatorial velocity multiplied by a mathematical function (the trigonometric sine) of the axial inclination to the line of sight, so that in any given case, all we can know is a lower limit to the rotation speed. In order to examine this characteristic for any class of star we must perform a statistical analysis on a large number of them under the assumption (long proven to be correct) of a random distribution of axial orientations. We then assume that the largest rotation velocities belong to those bodies whose axes are perpendicular to our direction of vision, and use this maximum to characterize the stellar type being studied.

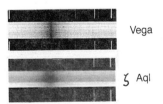

Figure 7.7. Stellar rotation rates are found from the broadening of spectrum lines, as illustrated by these two A0 dwarfs. Vega (top) has a rotation velocity (projected perpendicular to the line of sight) of only 15 km/s, and the Hδ line exhibits a sharp, well-defined center. Zeta Aquilae, bottom, of the same spectral type, is spinning at 345 km/s, which causes Doppler shifts that wash out the line. Sharp-lined stars like Vega could be rapid rotators viewed nearly pole-on. National Optical Astronomy (Kitt Peak) Observatories spectrograms, courtesy of E. C. Olson.

The cause of the abrupt increase in spin speeds, called the *rotation break*, is believed to be the process of magnetic braking that we introduced in Section 6.5, and which affects only the later spectral types. Presumably, these cooler stars once rotated with the higher speeds of the earlier classes. However the magnetic fields that are produced by their deep convection layers have slowed them down. Since the size of the convection layer diminishes and eventually disappears as we proceed up the main sequence, so does the braking effect, resulting in stars that are spinning much as they were when they were born.

Support for this hypothesis comes from the measurement of rotation speeds of stars in galactic clusters. In young assemblies, those that have their O and B main sequences still intact, all the stars spin with roughly the same periods, so that there is no sharp rotation break. The later types turn with somewhat lower speeds only because their dimensions are smaller. But as we look to older clusters, whose higher mass stars are missing and whose ages more match those of the unattached field (non-cluster) stars, we see that the later types are now spinning distinctly more slowly: something is indeed braking them, most likely their own magnetism. The very rotation that helps produce the magnetic dynamo thus inhibits and limits itself at the cooler end of the main sequence.

7.4 The Hertzsprung gap and passages to the A stars

In class F the giants cease to be a separate branch truly distinct from the dwarfs. The absolute luminosities of class III stars reach a minimum at about F4, where they closely approach the main sequence. Toward higher temperatures, the loci of the two types of stars run roughly parallel to one another, separated by about one magnitude (see Figure 3.8). Near the convergence point, the late F and early G giants become remarkably rare. Here we find the famous Hertzsprung gap between the giants and dwarfs, where there is a severe paucity of stars. The gap is illustrated by the old HR diagram plotted in Figure 3.6; in more modern diagrams that show more stars we see that it broadens upward into the supergiants to span from about A5 to M0. Once again, the F types demonstrate stars in transition, as the gap is caused by particularly rapid stellar evolution; we will examine it again in the final chapter of this volume.

In addition, at class F we begin to see a remarkable group of 'chemically peculiar' main sequence stars, in which the abundances of specific, usually heavy, elements can be greatly enhanced. Various types extend into class B, and we will discuss them more fully under the A stars where they are most prevalent.

In Figure 3.10 we saw that at F8 there is a crossover in the color–temperature–spectral class relations that extends earlier to about A0. Here the luminous stars are hotter than the dwarfs of the same type rather than cooler (Section 3.7). This effect is primarily a result of the particular standards and the specific lines that are used to define class: that is, temperature is derived from a variety of criteria that only partly include the absorptions used to determine the type, so that we do not necessarily expect consistency. Remember that inherent to the basic idea of spectral classification is an independence from physical parameters.

Finally, the shrinkage of the outer envelope's convection zone compounds the lithium problem examined in Section 6.10. According to stellar structure theory the F0–F5 stars should have 'cosmic', undepleted Li abundances, similar to that of the interstellar gas from which they were born. Yet over half show severe depletion. We do not know what is happening to it, but we are concerned that a pernicious process called *diffusion*, of importance in type A, may be at work, wherein heavier atoms simply settle downward and out of the upper atmosphere. In that case we are not getting an accurate assessment of average atmospheric properties. We do not know whether or not the process can affect the compositions that we measure in cooler stars. Obviously, our theories need improvement.

7.5 Instability: Cepheid variables

With class F, we enter the true heartland of the variable stars. Our HR diagram now reveals a large portion of a vast *instability strip* that was only hinted at in the last chapter. The zone extends from bright G supergiants all the way down through early F and A giants, subgiants, and dwarfs. Although the varied groupings and types of stars within the strip may have quite different characteristics all the styles of variation have a common underlying cause: pulsation. The stars change their radii, and consequently their temperatures and magnitudes, in a remarkably regular pattern, driven by an unstable zone not too far below the surface.

Of the different kinds of regular pulsators, the best known must be the Cepheid variables, named after the third magnitude prototype δ Cephei. Several other naked eye stars are also well known members of the class, η Aquilae and Polaris, for example (Table 7.1). They are actually relatively rare, but since they are typically supergiants they can be seen for great distances, so we observe a very large number. The stars' light curves (the graphs of magnitude plotted against time; see Figure 7.8) are characterized by a rapid rise followed by a slower decay, and the most remarkable characteristic must

Table 7.1. *Four bright Cepheids*

The reader might attempt to calculate the distances from the period luminosity relation and the average apparent magnitude. But be aware that the last three Cepheids are dimmed and reddened somewhat by interstellar dust, and that the distances will be lower limits unless the absorption is taken into account through a comparison of the star's observed color with that expected from its spectral type. We will look further at such a procedure in Chapter 9 when we encounter the topic of interstellar dust.

Star	Period (days)	Visual magnitude	Spectral type
Polaris	3.97	1.94–2.05	F7 Ib–II
δ Cep	5.37	3.90–5.09	F5 Ib to G2 Ib
η Aql	7.18	4.08–5.36	F6 Ib to G2 Ib
ζ Gem	10.15	3.68–4.16	F7 Ib to G3 Ib

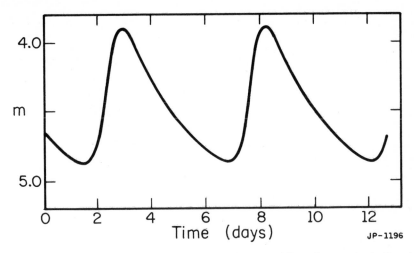

Figure 7.8. The light curve of δ Cephei. Adapted from *The Story of Variable Stars*, by L. Campbell and L. Jacchia, Blakiston, Philadelphia, 1941.

be their regularity: the plot of magnitude versus time is repeated over and over, with practically no change from one year, or even from one century, to the next. Amplitudes (differences between maximum and minimum magnitudes) for various stars are typically one-half to two magnitudes, and the periods range roughly from 2 to 100 days. The pulsating nature of these stars is proven by regular Doppler variations in the spectrum lines that demonstrate alternating expansion and contraction. Correlation with the light curves show that Cepheids are brightest not when they are largest, but at the time of maximum outward velocity.

Certainly the most interesting property of the Cepheids was discovered in 1912 by Henrietta Leavitt of the Harvard College Observatory. In a study of the Small Magellanic Cloud (Figures 1.7 and 7.1), one of our small companion galaxies, she found that the apparent magnitudes of the Cepheids correlate with their periods, with the slowest pulsators the brightest. Since all these stars are about at the same distance, that means that the true luminosities correlate with periods. Physically, the relation is not difficult to understand: the more luminous stars must be the more massive and therefore the larger, and their very size requires longer pulsation periods.

We thus have an obvious and very powerful way of deriving distances. If we can calibrate the Cepheid period–apparent luminosity diagram of the Magellanic Clouds by finding the distance and absolute magnitude of some Cepheids in our own Galaxy, we can derive the relation between absolute magnitude and period (Figure 7.9). Then, wherever we find a Cepheid, we can know how far away it is. The period gives us the absolute magnitude, and comparison with the apparent brightness yields the distance from the magnitude equation. The Cepheids are so luminous that they can be seen individually enormously far away, and are easily recognizable in other nearby galaxies

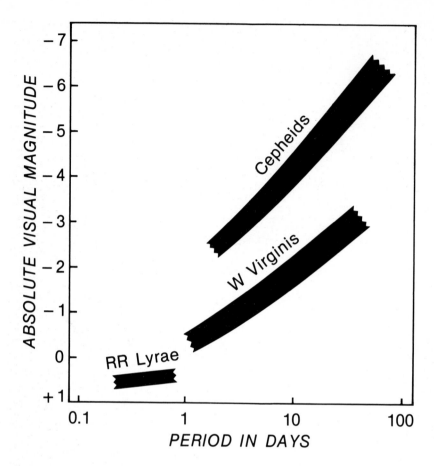

Figure 7.9. The period–luminosity diagram for the two populations of Cepheids, with the RR Lyrae stars at lower left. Compiled with data taken from *Astrophysical Quantities*, by C. W. Allen, Athlone Press, London, 1973.

(Figure 7.10). Thus they provide us with one of our best tools for mapping the nearby Universe in depth.

But, the problem is not quite that straightforward. Nature has given us two kinds of Cepheids, a set belonging to each of the two stellar population types (Section 1.8). Those of Population I, the set of celestial objects that make up the galactic disk, are more common and are referred to as *classical Cepheids*. The Population II, or galactic halo, variety are often called *W Virginis Stars*. Both follow a period–luminosity correlation, but for a given period, the classical version is about two magnitudes brighter than the other (again, see Figure 7.9). This separation is caused by a combination of differences in mass and chemical composition. The disk Cepheids, with masses five to fifteen times solar, have some ten times the bulk of the W Vir stars. The lower mass in itself should markedly decrease the luminosity. But the depressed metal content that is so characteristic of Population II *raises* it, although not fully compensating for the large mass

Figure 7.10. Cepheids discovered by E. P. Hubble (the numbered stars) in the nearby spiral galaxy M33, 1.2 mega (million) parsecs away. The lettered stars surrounding the variables are comparisons used for magnitude determinations. The bright image labeled HS 'B' is one of the supergiant Hubble–Sandage variables. Palomar Observatory photograph taken by W. Baade, from an article by A. Sandage in the *Astronomical Journal*. Print courtesy of Dr. Sandage.

difference. The metal atoms in stars within the temperature range discussed here are the primary sources of the star's *opacity*, the quality of a gas that impedes the flow of radiation. The fewer metals there are, the more easily the energy can escape, and the brighter the stars will be.

In order to derive correct distances, it is clear that period measurement alone is insufficient: we must also properly classify the star. If we mistakenly assume that a W Vir star is a Population I Cepheid, we will place it twice as far away as we ought. Light curves have been used for such discriminations, as those of the W Vir stars more commonly exhibit a bump or pause on their descending slopes, but the differences are small enough to lead to confusion. The most certain way of culling the Population II variety is to look at the metal content, which can be derived from the stars' spectra.

An additional, although far less serious complication to distance measurement, is the existence of a spread in the absolute magnitudes for a given period. We have found that this dispersion correlates with spectral type, and that what we truly have is a period–luminosity–color diagram. Obviously, to derive any accurate distance, the observer must have a considerable amount of information about the star.

7.6 RR Lyrae stars

In accord with the above discussion, we see on the HR diagram that the Population II Cepheids are located down the instability strip relative to their classical counterparts. They seem to be some sort of cousin to the luminous, longer period RV Tauri stars that we introduced in the last chapter, but the exact relation is not clear. Both occur in globular clusters. If we now proceed downward in the diagram, we encounter other kinds of Population II pulsators. Just below absolute magnitude zero, in early F, we find the *RR Lyrae* stars, which extend into late A. They abound in globular clusters, and are sometimes called 'cluster variables'; they are second in number known behind only the long period variables. Because they are all roughly at the same absolute magnitude, around +0.5, like the Cepheids these stars are excellent distance indicators, except that they cannot be seen as far. The brightest is some 7^m fainter than the most luminous Cepheid, and therefore the extreme limit of visibility is 25 times smaller.

The RR Lyrae stars historically played a powerful role in the discrimination of the two types of Cepheids, and in establishing the modern distance scale of the Universe. Prior to 1950, it was believed that the Cepheids were a homogenous group, and were roughly contiguous with the RR Lyrae stars on the period–luminosity diagram; we thought them to be much fainter than they actually are. When the 200-inch telescope went into operation in the late 1940s, however, the astronomers found that they were not able to see the RR Lyrae stars in the Andromeda Galaxy, M31 (Figure 1.9), as expected. At the distance inferred from the Cepheids, these cluster variables should have been readily visible. This inconsistency led to several other investigations that culminated in the realizations that there were two different kinds of Cepheids, that the period–luminosity diagram then in use was 1.5 magnitudes too faint, and that we were grossly underestimating the distance to M31 and consequently to every other galaxy.

Our concept of the size, as well as the age, of the Universe dramatically increased literally almost overnight. The reader should be well aware of this discovery in reading any astronomy book written before 1952.

7.7 Cepheid calibration

The above problem well illustrates the difficulty of the Cepheid calibration. The matter is far from simple. Cepheids are rare and distant, well outside the local zone where we can determine parallax (Section 5.7). Early attempts at calibration simply were not good enough to determine the true absolute magnitudes of the disk Cepheids, which resulted in the false merger of those of the two population groups. Even now, our data are not entirely satisfactory, and new studies are periodically undertaken to improve our knowledge of the galactic pulsators. The most obvious method is to infer the distances of a handful from their memberships in galactic clusters. The cluster distances, in turn, are found by comparing their main sequences with that of the Hyades, whose distance is known to high precision (Section 5.7). The method is compromised by the small number of Cepheids available and by the dimming effect of interstellar dust, which makes the clusters look farther away than they really are (we will explore this matter again in Chapter 9).

Another useful procedure is that of *statistical parallaxes*, wherein we can determine the average distance, and consequently the mean luminosity of a homogeneous set of stars. The basic assumption is that the motions of the stars are truly random, that is, there are no systematic effects that can cause the stars to move in a preferential direction. We first adopt our sample of stars spread out over the whole sky, and determine radial velocities and proper motions for all of them. Then we determine how the Sun is moving relative to the lot of them, that is, we find the solar velocity (speed and direction) required to make the *average* stellar motion equal to zero (so that the individual movements are random).

Now, we have an effect somewhat similar to parallax. We are on a moving platform, and we should be able to see the stars move past us as we plow through the group. The farther a star is away, the less will be its proper motion, and if we assume that it is at rest relative to the whole set, we can determine its distance from how much it appears to move over the course of a year. (As you drive on a roadway, a distant tree will appear to move across your line of sight more slowly than a cow at a nearby fence). But, the star certainly has its own motion that is peculiar to it alone (the cow may be walking, which distorts its relative movement), which will render any such calculation of distance erroneous. If however, we assume that the *average* motion of *all* the stars is indeed zero, then the average distance of the stars individually calculated under the assumption of zero speed will be correct. We perform these calculations on a set of Cepheids with common periods and thereby determine their mean absolute magnitude. Although more stars are available this way, the method does still present serious problems, notably interstellar absorption (which can produce erroneous luminosities), small and hard-to-measure proper motions, and (most importantly) that there may well be systematic effects in the motions.

Of course we could also simply determine the distances to the Magellanic Clouds (and their Cepheids) directly by employing other indicators. For example, we can find the luminosities of galactic O and B stars from the usual main-sequence fitting, and then use the apparent magnitudes of those in the Clouds to determine how far away they are. But then we have to account for the effects produced by the cloud's lowered metallic compositions. Nevertheless, even with all the problems, we have actually come up with a reasonable calibration, one that serves us well.

7.8 Fainter pulsators

We are still not near the bottom of the instability strip. Below the RR Lyrae stars, cutting through the giants of classes A and F, across the sub-giants, and even a bit into the main sequence, we find a large number of *dwarf Cepheids*, sometimes called δ *Scuti*, or more rarely, *AI Velorum* stars. The latter are those with generally larger variations. Consistent with their low masses and radii, they have short periods typically of a few hours, and small amplitudes of a few tenths of a magnitude. The brightest is ϱ Puppis, which varies between magnitudes 2.68 and 2.78 over a 3.4 hour period.

And we are *still* not done. The strip actually extends all the way down to the white dwarfs (Section 5.8), where it crosses over into the temperature region appropriate to main sequence spectral class B. We will look at some of these odd pulsators, with periods measured in minutes, later on.

7.9 Origin of the instability strip

Why is there such a thing as an instability strip? What makes these F types so unstable? Most stars are in a state of what we call 'hydrostatic equilibrium', in which for any stellar layer, the outward forces of gas and radiation pressure just balance the inward pull of gravity. Although photons carry no mass, radiation does exert a pressure. It is insignificant on Earth where the flow of light is low, but in a star there is so much that its push becomes a important factor in stellar structure. It in fact limits the ultimate masses of stars and drives stellar winds.

The Cepheids and their kindred have lost this balance: the density within some of the interior layers is first too high, and the star must expand with a drop in pressure, which throws it past the equilibrium point. Gravity eventually takes the upper hand, stops the expansion, and initiates contraction. The pressure then builds to the point where it wins out over gravity, the compression halts, and the process repeats itself.

The culprits behind this behavior are the star's hydrogen and helium. At the surface, where the temperature is low, both kinds of atoms are neutral. As we penetrate inward, and the temperature increases, we first pass layers in which the H and He atoms are becoming singly ionized, and then a deeper zone where the other helium electron becomes stripped away. Normally, as you increase the temperature of the gas in a star, its opacity decreases. But in these regions where the act of ionization is taking place, the gas behaves oppositely. If you give it a squeeze, both pressure and temperature rise and the increased opacity blocks the radiation, which then gives the layers an outward push. That results in a lowering of the density, temperature and opacity so that the outward shove of radiation is diminished, and the layer relaxes backward producing another squeeze. The ionization layers then are seen to act as a radiation valve that drives a continuous oscillation.

Because even the second helium ionization zone is not too far down, Cepheid oscillation is basically a surface phenomenon that extends only about 20% of the way into the star, and involves only a small fraction of its mass. The stellar interior produces its energy at a steady rate, and knows nothing of the variability. In order for the mechanism to work, these driving zones must be at just the right depth so that they have sufficient density and mass to cause the outer parts of the star to pulsate, but not so deep that the effect cannot be felt at the surface. The proper conditions exist within this famous strip that passes from the luminous G supergiants through the heart of class F, down across the main sequence (where its effects are considerably suppressed) and below into the white dwarfs. In the other direction the driving mechanism extends its effects into the region of the RV Tauri stars and into the realm of the red giants to produce the Mira variables.

As a coda to this section, note in the accompanying HR diagram that the prominent star α Aquarii falls right in the middle of the strip, placed there strictly on the basis of its classification as a G2 Ib supergiant. It is most certainly *not* a Cepheid, and its location illustrates the uncertainties inherent in classification.

7.10 Brightest to dimmest

The instability strip stops above absolute magnitude minus six. Beyond this limit, many of the stars are variable in the erratic way that frequently afflicts the massive, huge and thus highly attenuated supergiants. Included among the most luminous F stars is one of the heavens' most bizarre objects, R Coronae Borealis (classified cFpep: even its emission lines are peculiar in nature). This star is normally of apparent magnitude 5.8, and can easily be seen inside the Crown's half-circlet. But every few years it suddenly plunges nine or so magnitudes fainter, as it becomes enshrouded in a dust cloud produced within its strong outflowing wind. About 40 similar stars, most commonly F and G supergiants, are known; a few exhibit Cepheid properties as well. The cause of the phenomenon is not known.

We see a wonderful collection of other strange stars related to this class. FU Orionis, for example, is currently a late F or early G supergiant roughly of luminosity class II. It, V1057 Cygni, and a few others represent a class characterized by a rapid (year long) brightening of five or six magnitudes followed by constancy or a slow decline. FU Ori, the prototype, has stayed in its high state for nearly 50 years. They seem somehow to be related to the T Tauri stars (Section 5.6) and to stellar birth. As another example, the great star clouds of the southern Milky Way contain one of the oddest of all stars, η Carinae. In the nineteenth century it was one of the sky's brightest stars; it has since faded to naked eye invisibility and its once F-type spectrum has been replaced by one of pure emission. We will look at it again in Chapter 11. And do not forget ε Aur, the strange eclipsing binary that we examined in Section 5.4.

Outside of our own Galaxy we find some equally odd beasts. The photo of M33

(Figure 7.10) with its many Cepheids also shows a *Hubble–Sandage* (*HS*) variable. These brilliant stars, near absolute magnitude −9, rivalling ϱ Cas, have middle-F spectra. HS variable A in M33 (not indicated) has faded dramatically and now appears as type M. These peculiar erratics appear to be expelling matter, and the spectrum is formed by the circumstellar shell rather than by the star itself buried within.

From ϱ Cas at the very top of the diagram, we now take a precipitous drop of 23 magnitudes – a factor of over one *billion* in luminosity – to visit briefly again with the white dwarfs, and to return to Procyon, the star with which we opened our discussion. This first-magnitude F subgiant has an 11th magnitude companion only a few seconds of arc away from it. Procyon B has not been well studied, but its temperature seems to be roughly appropriate to that of a late A star, not too much hotter than Procyon itself. The difference in brightness can only be due to a difference in surface area or radius: the companion must only be about twice the diameter of the Earth. We can observe the orbital motions of the components of Procyon, and from Kepler's third law (Section 1.6) we find, quite remarkably, that Procyon B has a mass over half that of the Sun. It must thus be incredibly compressed and terribly dense, on the average of a million times that of water, or about a metric ton per cubic centimeter.

We have met the white dwarfs so far but once, when we encountered them among the K stars (Section 5.8). We shall treat them much more fully when we examine the prototype, the companion of Sirius. Curiously, each component of Procyon has now announced its Sirian analogue. And now that our transition is complete, we proceed to the three classes of hot stars: A, B, then O.

8 *The white stars of class A*

Now that brilliant Sirius, with its white dwarf neighbor, has been properly announced, let us look at it and its great company of A stars. None – of any class – dominates the sky like α Canis Majoris, the Dog Star, Orion's companion, seen within its constellation in Figure 8.1. At apparent visual magnitude -1.46, it is $0^m\!.7$ brighter than number two, Canopus. And for most residents of the northern hemisphere, who can see neither α Carinae (Canopus' other name), nor α Cenauri (number three), it appears $1^m\!.4$ more luminous than its nearest visible rival, Arcturus. Sirius (A1 V) is so lustrous that it can be seen through the window of a modestly lit room. Its name alone, which means 'searing' or 'scorching', attests to its brilliance. Although this star is truly luminous, intrinsically almost 50 times brighter than the Sun, its fame is more the result of proximity. It is the fifth nearest to us – counting multiples as single units – with a gigantic parallax of $0''\!.375$, which yields a distance of only 2.7 parsecs or 8.7 light-years. A true mark of northern hemisphere Winter, α CMa culminates the meridian at 8 p.m. in early January, when the notorious refraction of clear, cold skies makes its white light sparkle with countless colors. When northerners finally see it slipping away into the western twilight, they know that the year's chill is once again retreating.

The naked-eye sky is quite well populated with the As. The count of stars per unit volume of space drops precipitously along the main sequence toward earlier types, but increasing luminosity renders this class very visible, and their number consequently appears to increase. Of the 22 stars of the first (or lower) magnitude, five are of type A, with four of them – Vega, Sirius, Fomalhaut, and Altair – neatly delineating the dwarf locus from A0 to A7 (Figure 8.2).

A remarkable cache of A stars is contained in the best known figure of the northern sky, the Big Dipper. The five central members, β through ζ, are all on the main sequence between A0 and A3. Their similar apparent brightness shows that they are all at about the same distance from us; (excluding ε UMa, which is farther away, they average about 25 pc). As pointed out before (Section 1.7), they are part of a poorly populated moving galactic cluster, which also includes Sirius, α Coronae Borealis (another A0 V star), and several others farther down the main sequence.

Among our first magnitude stars we find one other of this class, Deneb, our much-used example in Chapter 1 and an A2 Ia supergiant. This magnificent star, simultaneously the tail of the Swan and the head of the Northern Cross, shines at

Figure 8.1. Canis Major and Sirius. The star at the right-hand edge is β CMa, Murzim (B1 II–III), who announces Sirius' rising. The galactic cluster M41 is directly below Sirius; several others can be seen in the part of the Milky way that runs through the photo. From *An Atlas of the Milky Way*, by F. E. Ross and M. R. Calvert, University of Chicago Press, Chicago, 1934.

Figure 8.2. The HR diagram for classes A and F. The position of the horizontal branch is indicated by the placement of RR Lyrae at F2. Vega and Sirius shine brilliantly here; Deneb is visually fainter because it is so much farther away. Note how the white dwarfs define a tight sequence. These tiny stars are placed according to their temperatures and the main sequence temperature scale, not by the spectrum lines that appear. See Figures 3.8 and 12.7 for credits.

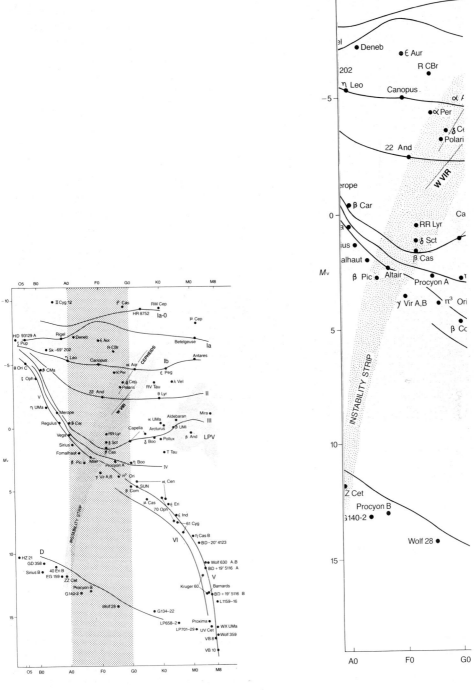

Figure 8.2.

apparent magnitude 1.3 even though it is some 500 parsecs away. Its absolute visual magnitude of −7.3 makes it about the most intrinsically luminous A star known in our Galaxy. Its radius of 50 Suns would overflow the orbit of Mercury. The A supergiants are really quite rare. Only one other, η Leonis (A0 Ib), just north of Regulus in Leo's sickle, is third magnitude or brighter; and fainter in absolute magnitude, we find absolutely *no* naked eye stars in MKK's luminosity class II.

The giant region is more populated. Here we find such well-known stars as Rasalhague (α Ophiuchi, A5 III), γ Bootis (A7 III), and 'Orion's Footstool', β Eridani (A3 III), the wellspring of the celestial river, which ends in that southern luminary, Achernar (B5 IV).

8.1 Spectra

Recall that the original Harvard classification scheme of 1891 was based largely upon the hydrogen lines, and by definition it is at class A that they reach their greatest strength (Figure 8.3). We have watched them climbing out of the background since early M, and now their domination is total. If there is any spectral class that is instantly recognizable on even the poorest quality and lowest dispersion spectrograms, it is A.

The neutral and even singly ionized metal lines are fading fast. The huge number of iron and titanium absorptions, which with others give the spectrum a broken,

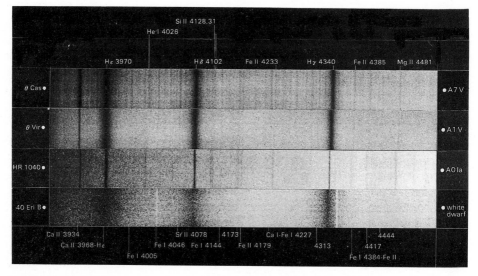

Figure 8.3. A selection of spectra. They are not quite at the same scale, which results in a slight offset, particularly noticeable at λ4481. The He I and Si II identifications at the top apply only to the supergiant. The absorptions in the white dwarf are all hydrogen. The bright lines are caused by mercury street lamps near the observatory. Note the great strength of the hydrogen lines throughout this class, and the disappearance of the metal lines toward earlier sub-types. The hydrogen lines become quite narrow in the spectra of the rarefied supergiants, and remarkably broad in those of the dense white dwarfs, a sure guide to luminosity. In addition, the metal lines return faintly for the supergiant. Compare with Figure 3.7. From *An Atlas of Representative Spectra*, by Y. Yamashita, K. Nariai, and Y. Norimoto, University of Tokyo Press, Tokyo; John Wiley & Sons, New York, 1978.

Figure 8.4. The spectrum of the classic A0 star Vega between λ3640 and λ4585 Å. The broad prominent lines are all hydrogen. Note their steady progression into the ultraviolet until they overlap just longward of the Balmer limit. The H line of Ca II is within Hε, just to the left of the sharp hydrogen line core. Most of the other lines faintly visible are iron. National Optical Astronomy (Kitt Peak) Observatories spectrogram, courtesy of E. C. Olson.

chopped appearance, are still faintly visible (especially at the higher dispersion of Figure 8.4) but are weakening with increasing temperature. By A0 even the once powerful Ca II lines are but a vestige of their former selves, and at the dispersions used for classification, Ca I at λ4226 has completely disappeared. Thus the optical spectrum assumes a simpler appearance, broken at first glance only by the stately progression of Balmer lines as they proceed toward the second level ionization limit at λ3646 Å (see Section 2.4). The simplicity of the hydrogen spectrum displayed so beautifully here in Figure 8.4, was instrumental in the discovery of the nature of the atom and of the structure of electron orbits, and in the establishment of the science of quantum mechanics in the early part of this century.

The hydrogen lines remain relatively steady in strength throughout this class, with a broad maximum at A2. The Ca II K line is rapidly diminishing toward the hotter types (Fraunhofer H of Ca II is coincidentally blended with the Hε line at λ3970): consequently, the comparison of calcium K to Hε provides a very sensitive classification indicator. The rapid change with temperature seen for λ4226 of Ca I makes it useful as well, and it is often compared with the relatively steadier Mg II absorption at λ4481.

Class A presents us with the most dramatic example (used in Chapter 3) of spectral change with variation in absolute magnitude. In types K and M, the hydrogen lines show us a positive luminosity effect (i.e. they strengthen as the stars brighten), which disappears in classes G and F. Now, their behavior reverses to a negative effect, which causes hydrogen absorptions to be much weaker in supergiant than in dwarf spectra (Figure 8.3; see also Figure 3.7). The cause is readily explainable in terms of atmospheric pressure. As we found in Section 2.8, all spectrum lines have a certain width in wavelength, part of which is caused by the perturbations of the electronic energy levels from their normal positions under the action of atomic collisions. These levels then become somewhat smeared out, and that allows an electron to absorb light slightly away from the line center. Consequently, the higher the pressure, the broader the line. At the low densities found in supergiants such collisions are relatively infrequent, and the hydrogen lines are naturally much narrower. At the other extreme, at the extraordinary densities of white dwarfs, the absorptions attain amazing breadth. In the late (K and M) stars, these lines are so intrinsically weak that this pressure effect is not noticeable, being overridden by more important factors such as differences in atmospheric opacity.

The giants of class A are only a magnitude or so brighter than the dwarfs, and only about 50% larger. The Balmer line widths are therefore not a very good discriminator among classes III, IV, and V, and we must rely on subtle positive luminosity effects exhibited by the spectra of Fe II and Si II.

8.2 Line structure and abundances

The superb high dispersion spectrograms of brilliant Sirius presented here (Figure 8.5), made with the coudé spectrograph of the Mt. Wilson 100-inch telescope, illustrate some of the detail of the construction of spectrum lines, and provide us with an opportunity to explore the formation of the absorptions and the calculation of abundances in greater depth. They are not simple sharp pieces cut out of the spectrum, made as if snipped with a pair of scissors, but show considerable structure. The great widths of the hydrogen lines relative to those of the metals is of course obvious. The metallic lines have some spread to them too, here caused mostly by stellar rotation (Section 7.3), which is especially noticeable on the lower, very high dispersion ultraviolet spectrogram. It is interesting to compare the upper spectrum in Figure 8.5

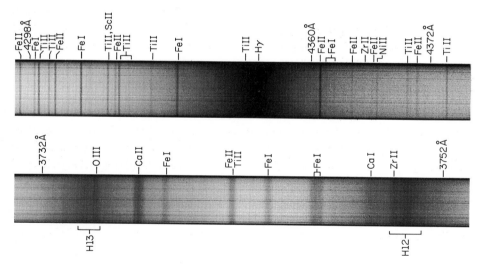

Figure 8.5. Two high dispersion spectra of the famous star Sirius, taken with the coudé spectrograph of the 100-inch Mt. Wilson telescope. The top print spans only 84 Å around Hγ, and covers the region 4295 Å to 4379 Å. Here we can see the enormous contrast between the narrow metal lines and the broad Hγ feature, which spans some 15 Å. Most of the metallic features are from singly ionized iron and titanium, although there are a pair of rather strong neutral iron lines, and a wisp of ionized zirconium. The bottom print is nearly four times the dispersion of the upper, and shows the region from 3730 Å to 3753 Å – a mere 23 Å – in great detail. Here we see the broad depressions caused by the H 12 and H 13 lines (produced by the 2 → 12 and 2 → 13 orbital transitions of hydrogen), a variety of metallic lines, and one of oxygen. Look at the structure of the calcium line, which features a 'core' flanked by more diffuse 'wings'. Mt. Wilson and Las Campanas Observatories, Carnegie Institution of Washington photographs, courtesy of R. and R. Griffin.

with those of the F5 stars Procyon and α Persei that appeared in the last chapter (Figure 7.6): note the changes that have taken place in Hγ and in the metal lines.

More to the point, however, look at the ionized calcium line in Figure 8.5 just to the right of H13, and you see a feature that is nearly twice as wide as the iron absorptions, and sports a sharp central *core* flanked by broader *wings*. This formation can also be faintly seen in the Ca I line to the left of H12, and in Hγ.

If we could experiment on a star and add atoms of an element to its atmosphere, we could watch a line grow in strength and width. It would first appear merely as a weak depression of the continuum. The feature would have a certain width to it that depends upon the rotation of the star (as for Sirius), the degree of turbulence (as for α Persei, Figure 7.6), the atmospheric pressure (Figure 8.3), and quantum mechanical parameters (Section 2.8). As we add more atoms to the gas, the line extracts additional light from the continuum, and it becomes darker, or deepens, but more or less maintains its width. With increasing abundance of the element, the center of the line becomes quite opaque, and the atoms begin to absorb a significant amount of light away from the middle, causing the growth of the wings. Once the core is completely black, or 'saturated', the only place the line can grow is in these flanks, and the wings become ever more prominent, as we can see in Hγ. Unfortunately, we cannot so experiment, but we can get the same effect by comparing a variety of lines of the same element that have different probabilities of formation.

A common way of measuring the total strength of a line – the amount of energy extracted from the background continuum – is through its *equivalent width*. We replace the real line with its core and wings with one that has the same absorbing power, but which has sharp edges that drop immediately to zero intensity. The equivalent width is then the width of this false line in Ångstroms, or milliÅngstroms (mÅ). A graph of the strength of the line plotted against the number of atoms acting to absorb it is called a *curve of growth*. It starts as a straight line, as one might expect, with the equivalent width increasing in concert with the number of absorbing atoms. But then it becomes more complex in response to saturation effects. For any ion, we can construct an effective curve of growth by plotting the equivalent widths of all the observed lines of an ion against the atomic probability of formation of the line. Then a comparison of curves of growth of two different ions will give us the ratio of the abundance of one relative to the other. If we so examine all the ions present for the various atoms, we can establish a reasonably accurate picture of the chemical composition of the star's atmosphere.

But that procedure just provides us with a rough estimate. The next step is to build a model in a computer that actually best matches the observed spectrum with all its variety (see Section 4.3). That is, we conceive of a radiating stellar atmosphere that has a certain temperature and pressure structure, calculate how the electrons are distributed among atomic energy levels or orbits, and then compute what the strengths, widths, and detailed structures of the various lines should be. We finish by adjusting the relative abundances of ions, as well as the physical make-up of the atmosphere, until we get the best possible fit between theory and observation. This more sophisticated approach requires a vast amount of spectrographic data, a great deal of physical knowledge about the structures of atoms, and considerable computer expertise. The final result, whether by crude estimate or detailed modeling, is that we have been able to assess the myriad chemical differences among stars that range from the gross distinctions among M, R, N, and S stars to such specifics as the intriguing variety seen within class A discussed below.

8.3 Strange dwarfs: the metallic-line stars

The majority of stars on the A main sequence show simply the normal progression from F to B, becoming more massive and luminous, and larger. Vega, for example, radiating at 55 solar luminosities, has three times the Sun's mass and diameter. Average rotation speeds steadily increase from about 100 km/s at F0 to about 180 at A0 on their way to their maximum among the B stars. The X-rays that are observed farther down the main sequence are not generally seen earlier than about A5 as the convective layer responsible for dynamo action, magnetic fields, and solar-type activity completely disappears as a result of changes in the stars' internal structures.

What is surprising, however, is the large number of stars – perhaps as many as 25% – that have odd spectra, with unusual elemental line strengths and other peculiarities (Figure 8.6). We begin to see these at mid-F. Their ranks extend to early

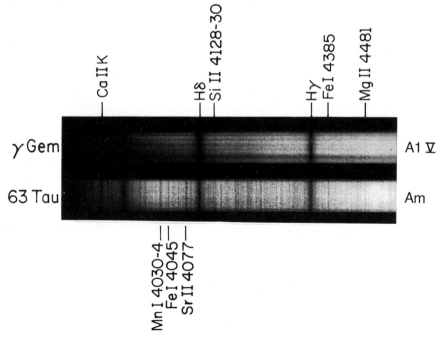

Figure 8.6. The classic Am star 63 Tauri, compared to the normal A1 V dwarf γ Geminorum. They have the same Ca II K-line strengths, yet the general metallic spectrum is anomalously strong in 63 Tau, more typical of the cooler F stars. From *An Atlas of Stellar Spectra*, by W. W. Morgan, P. C. Keenan, and E. Kellman, University of Chicago Press, Chicago, 19743.

B, and they attain their greatest number at A. The *chemically peculiar*, or *CP* stars as they are sometimes known, comprise about a half-dozen distinct groups that are related to effective temperature. Here in type A we encounter the two most famous of them, the *metallic line* (*Am*) stars and the *peculiar* (*Ap*) variety (Section 8.4) whose stars frequently possess powerful magnetic fields.

The Am stars have calcium and scandium lines that are too weak, and heavy metal absorptions (beyond iron in atomic number) that are too strong. These include copper, zinc, strontium, yttrium, zirconium, barium, and the rare earths: that group of odd metallic elements that lie between atomic numbers 57 and 71. The unusual line strengths can make classification difficult. The weakness of the K line makes the star look too early for its temperature, and strengths of the ionized metals can make it look too late. This stellar variety actually starts at around F4 (where they are called Fm stars) and extends upward in temperature to about A0. The differences in their spectra as compared with those of normal dwarfs are sometimes subtle, however, and are not always visible at classification dispersions. Sirius itself is a mild Am star.

A general characteristic of the Am group is that they are unusually slow rotators. As we saw in Section 7.3, the line widths (after accounting for other broadening effects) are used to derive rotation speeds relative to the line of sight. In order to account for the effect of random axial inclinations in space, we must analyze a large statistical sample of the spectra of these stars relative to those of type A in general. The maximum spin rates we find are not very large. Thus we assume that Am stars in general rotate quite slowly.

In addition, a high proportion of these odd objects reside in close binaries. When two stars orbit near one another, each raises tides in its partner that gradually cause synchronous rotation, just like the behavior of our Moon relative to Earth. The tidal interaction appears to be a mechanism by which the slow rotation speeds that seem to be necessary for Am development can be produced: the same phenomena that speeds up the much more slowly spinning G stars to form the RS CVn variables (Section 6.11).

8.4 The magnetic peculiar A stars

Stranger still is the other group, the *peculiar A* or *Ap stars* (Figure 8.7), which are well known for their strong magnetic fields. (The exceptions are the *mercury–manganese stars*, which can also carry the Ap designation; most of these are in class B, however, and we shall look at them again in the next chapter). These start at about F0 and extend earlier to about B4, where they are, of course, called Bp. They too possess

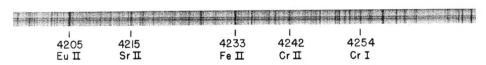

4205	4215	4233	4242	4254
Eu II	Sr II	Fe II	Cr II	Cr I

Figure 8.7. A high-dispersion spectrogram of the A0p star HD 125248, a 6th magnitude magnetic variable in Virgo, with a field strength about 4000 times that of Earth. Note the great strength of the europium line. From an article by H. W. Babcock, in *Stellar Atmospheres*, J. L. Greenstein, ed., University of Chicago Press, Chicago, 1960.

overabundances, but of different elements. Here, we find anomalously strong lines of silicon, chromium, strontium, and again the rare earths, most particularly of europium. The odd line strengths are rather easily seen, and consequently the stars carry the 'p' (peculiar) suffix attached to their spectral types.

The magnetic fields associated with these stars were first found in 1947 for the A2p star 78 Virginis. Stellar magnetism is analyzed via the *Zeeman effect*, in which spectrum lines are split into multiple components. In each of these the light is *polarized* in a characteristic way. In perfectly polarized light the waves all oscillate in the same direction; partial polarization, in which there is a preferential direction, is common in nature and is easily detected. We will explore the subject further in Section 9.9. Because of a variety of broadening effects, the actual splitting in stellar spectra cannot usually be seen, but the changes in polarization across a line can be detected, from which the magnetic field strength can be inferred. There is also some evidence for X-ray production among the Ap stars, which may be related to the magnetic fields, but their origin is not understood.

The magnetic fields can be extremely strong. In the above example, H. W. Babcock found a field strength of 1500 gauss; by way of comparison, the Sun's general field, outside of the active areas, is only about one or two gauss. The strongest known among this type (for which the field is exceeded only by those of the magnetic white dwarfs, which we will examine later) belongs to ninth magnitude HD 215441 (A0p), with an astounding 34 000 gauss field. The best known is probably the bright member of Cor Caroli (α^2 Canum Venaticorum, A4p), which measures in at around 1000 gauss.

The fields are almost always regularly variable, and we sometimes see complete reversals of polarity (field direction). Variations in magnitude and line strengths accompany the changes in magnetism. In order to explain the observations, the stars almost have to be *oblique rotators*, in which the magnetic and rotation axes are not aligned. This phenomenon is common: Jupiter, Uranus, and the Earth possess such inclined axes. The field strength we see depends upon the orientation of the magnetic

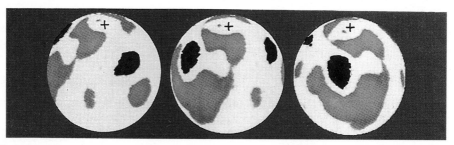

Figure 8.8. Three views of the reconstructed surface of the A1p Si star γ^2 Arietis. The 'Si' suffix shows that the spectrum has overly strong silicon lines. The changes in the Doppler profiles of the lines (see the text) allows us to map the positions of enhanced as well as depleted silicon abundances, and presumably the regions of strong magnetism. Dark areas show zones of enriched Si, light areas depletions; the '+' marks the rotation pole. We can actually see the zones move across our line of sight as the star spins. Lick Observatory images, courtesy of S. Vogt, A. Hatzes, and G. Penrod.

axis to our line of sight, and if the inclination of the two axes has the proper value, we may see first one magnetic pole and then the other.

The areas of elemental overabundances do not seem to pervade the entire atmosphere as they do for the Am stars, but concentrate toward magnetic patches on the surface. Thus as the star rotates, zones with different composition are brought into view, and the line and magnetic strengths correlate. As the enhanced (and depleted) areas drift across the star's surface they will move through regions whose spectra are Doppler shifted by rotation first in one direction, then in another. The result is that the line profiles produced by the patches will vary with time. By observing the detailed changes during the rotation period we can actually reconstruct images of the star that show the locations of the active magnetic areas, as in Figure 8.8. The results of extensive studies show that the magnetic fields are more complicated than those produced by a simple bar magnet: no great surprise, considering the complexity of the Sun's.

8.5 Causes of the phenomena

We do not understand the origin of the fields. The Ap stars, like the Am variety, also tend to be slower rotators, with equatorial velocities under fifty or so kilometers per second. This broad relation at first suggests that the fields might indeed be generated by dynamo action and that the stars may have been magnetically braked. However, there are several serious problems with this notion. First, our data are not complete, since rotational line broadening in rapid rotators can hide the effects that might be produced by magnetism. Improving techniques do show that some fast rotators are in fact magnetic, and that if anything the more quickly spinning stars have the lower field strengths, counter to what we would expect from the dynamo mechanism. Then there is the problem of what could produce the dynamo action, since these stars lack the convective layer found among the later types. Finally, we do not know what to make of the many slow rotators that are not magnetic at all.

An alternative idea is that an Ap star carries a 'fossil' magnetic field left over from the time of its birth, which is strengthened as a result of the star's formation. However, how do we then explain the stars that exhibit no magnetic fields at all? It is also theoretically hard to see how a field could quietly be maintained for such a long time. Some kind of continuous generation seems necessary; but if not a dynamo, as understood in the Sun, what?

There are also a variety of ideas put forth to explain the Am–Ap abundance phenomena. One of the earlier ones invoked surface nuclear reactions that create certain elements at the expense of others. Another, not considered very likely, is the contamination of a stellar atmosphere from by-products of a nearby supernova. Another possibility that is seriously considered is the selective magnetic accretion of certain elements from the interstellar medium. Currently, the popular idea is diffusion: the process by which elements settle out of the atmosphere, or rise to higher layers, in response to effects of radiation pressure or magnetic fields. We encountered this concept first among the F stars (Section 7.4) where it appears to affect the abundance of lithium. The theories are complex and difficult, and none is at present entirely satisfactory in the interpretation of some of the very large overabundances observed.

Just to confuse the issue a bit further, we might also consider the rare λ Bootis stars. These are of Population I, but possess severe *underabundances* of the light metals. Perhaps some sort of surface diffusion is at work here as well.

8.6 Planetary disks?

We have long searched for evidence of planetary systems accompanying other stars. They have proved elusive, however, because of their low masses and feeble illumination. We have made some possible, though controversial detections of minor unseen companions that orbit nearby M dwarfs (Section 4.10), but even these, if confirmed, would considerably outweigh our own Jupiter. Our general belief that most stars may be attended by planets is derived from the observation that our Jovian planets have their own miniature satellite systems, and is inferred from the statistics of double star masses (i.e., Jupiter and the Sun are a sort of binary that represents an extension of the observed statistical distribution of mass ratios).

The A stars, however, are now providing us with a clearer view thanks to some remarkable technological feats. The first of this new information comes to us from the *Infrared Astronomical Satellite* (*IRAS*), an orbiting telescope that surveyed the sky in

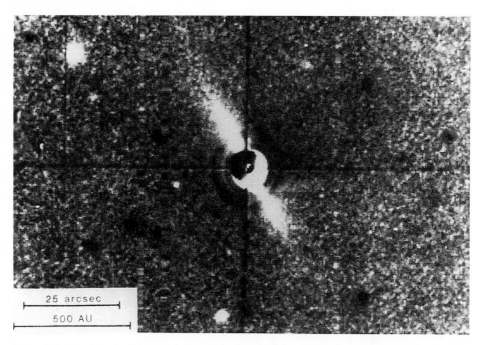

Figure 8.9. The disk of solid matter surrounding the star β Pictoris. The star itself is hidden behind an occulting disk set into the telescope so that it would not overwhelm the much fainter extended image. Are there planets buried within? Mt. Wilson and Las Campanas Observatories, Carnegie Institution of Washington, from an article in *Science*, by B. A. Smith and R. J. Terrile.

various infrared spectral bands during most of 1983. With this satellite, astronomers discovered a shell or disk of solid particles that surrounds brilliant Vega out to a distance of 85 AU. At a temperature of just under 100 K, the dust is quite invisible optically, but it radiates profusely at substantially longer wavelengths. The grains are estimated to be more than a millimeter across, and may represent the remnants of the formation of the star, and could be associated with much larger solids and even planets. Similar structures are found around Fomalhaut (A3 V) and the K2 dwarf ε Eridani.

More dramatic evidence of a solar-system-like structure is yielded by Earth-based observations of the southern A5 dwarf β Pictoris (Figure 8.9). With the advanced detector known as the charge-coupled device (the CCD; see Section 1.11) a pair of astronomers working at the Las Campanas Observatory in Chile detected a real disk of solids projecting outward to a remarkable distance of over 400 AU, ten times the dimension of our own known planetary system. Where there is such dust there could well be planets: witness the zodiacal light in Figure 1.7. We are really beginning to see into the local environments of stars in order to make comparisons with our solar home.

8.7 Instability

That great divider of the HR diagram, the instability strip, finally meets the main sequence full on at type A. We now can consider the complete range of the δ Scuti variables (Section 7.8), which extends along the dwarf locus from early A to F, and through the giants from late A to mid F. Approximately *one third* of the main sequence stars between A2 V and F0 V are low-level pulsators, some with amplitudes of only a few hundredths of a magnitude.

The qualifications for variability are only partly understood. Increased metallicity inhibits the phenomenon: the Am stars are generally of constant brightness. Among the δ Scuti giants, rapid rotation suppresses the pulsations. Members of this populous class sometimes also exhibit multiple periods. The subtle effects exhibited by these stars require immense amounts of observing time, which slows the acquisition of the data required to understand them.

8.8 The horizontal branch

Most of our time in this volume has been spent in looking at the classical zones in the HR diagram that pertain to the stars of the galactic disk, with a few excursions into the 'third dimension' of unusual abundances (Section 3.6), such as those discussed above, and those possessed by the subdwarfs and giants. of the galactic halo. Analysis of the stars of the archetypes of Population II, the globular clusters (Figure 8.10), reveals another major area, the *horizontal branch* (or *HB*) that cuts across the diagram at about absolute magnitude $+0.5$ between very roughly G0 and B2. It contains the RR Lyrae stars, which we first encountered in early F and which extend into late A (Section 7.6) in the zone crossed by the instability strip. The HB is quite prominent through class A, and crosses the main sequence at about A0 (Figure 8.11).

Although these stars may at this point superficially look like ordinary dwarfs, they are physically very different. Their energy comes largely from the nuclear burning of

Figure 8.10. The globular cluster M15 in Pegasus, which has a well developed blue horizontal branch and also, in contrast to the general rule, contains nearly 1000 RR Lyrae stars. In addition it is also one of two globulars known to contain a planetary nebula (Chapter 11), not seen at this resolution. University of Illinois Prairie Observatory photograph.

helium into *carbon* rather than hydrogen into helium (we will look at the details of this fusion in Chapter 12), and they are in quite an advanced state of evolution. In any case there is no ambiguity between the HB and the main sequence at type A, since the clusters, for evolutionary reasons described earlier (Sections 1.17 and 5.7), lack an upper main sequence. Massive stars always evolve first, and in these old clusters the dwarfs are present only later than mid F.

The HB is also found among the general, or field, population of halo stars, but we can still explore it best in the globulars. Astronomers divide it into three parts; the red and blue ends, and in the middle the *RR Lyrae gap* (Figure 8.11), in which one may or may not actually find these cluster variables. The gap extends, as noted above, from roughly F5 to late A. In spite of their homogeneous appearances, the globular clusters are actually a rather diverse lot. From one cluster to another, we see variations in the positioning of the giant branch, and in the distribution of stars within the HB. In some, like the great Hercules cluster M13, M22 in Sagittarius, and M15 in Pegasus (Figure 8.10), the blue side of the HB is highly populated relative to the red. In others, most prominently in the magnificent southern globular 47 Tucanae (seen in Figure 7.1), the red end is entirely dominant. Then there are those like M3 in Canes Venatici (Figure

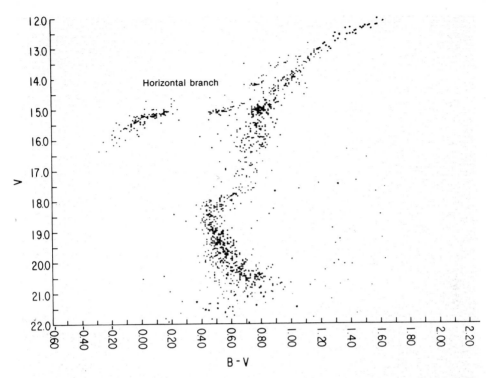

Figure 8.11. The HR diagram of the globular cluster M5 in Serpens, presented as a color–magnitude diagram. The conversion from spectral class to color is shown in Figures 3.10 and 3.11. This cluster has a fairly uniformly distributed horizontal branch and a well defined RR Lyrae gap at B−V about 0.3. Within the gap there are nearly 100 of these variables, which are not plotted. From a compilation of HR diagrams in *Dudley Observatory Report No. 11*, by A. G. D. Philip, M. F. Cullen, and R. E. White.

1.5) and M5 in Serpens Caput that take a middle ground with even distribution. It is in these latter ones that the RR Lyrae stars really abound: 100 are known in M5 and 200 in M3. The clusters with unbalanced horizontal branches, those with dominant red or blue ends, usually have far fewer; and the former are especially poor: 47 Tuc has but one RR Lyrae star, and some have none at all.

This distribution of stars along the HB is in part a result of metal abundance. In general, the weaker the metal lines, the more the HB extends into the blue, largely as a consequence of the difference in composition, which lowers the opacity of the stellar envelope through which the outgoing radiation must pass (and thus rendering the stars generally hotter). For example, M92 in Hercules, with a very blue HB, has an iron-to-hydrogen ratio only one percent that of the Sun, whereas in 47 Tuc it is reduced from solar by only a factor of three.

And here, we set the stage for one of the more intriguing problems in stellar astrophysics, epitomized by the clusters M13, M3, and NGC (the New General Catalogue of non-stellar objects) 7006. They have blue, intermediate, and red HBs respectively, yet all have similar metal abundances of about 3% of the solar value. Something other than metallicity also controls the HB distribution, but we do not yet

Figure 8.12. The nearly edge-on spiral M104 in Virgo. The fuzzy images around the periphery are globular clusters, from whose luminosities the distance can be derived. National Optical Astronomy (CTIO) Observatories photograph.

know what this *second parameter*, as it is referred to, might be. It may involve variations within the broad term of 'metallicity', such as differences in carbon, nitrogen, and oxygen abundances, or perhaps rotation.

The globulars are an excellent probe of galactic and intergalactic space. In the nearby clusters the brighter dwarfs are visible, and we can employ the usual method of main sequence fitting (Section 5.7) to determine their distances, which also establishes the luminosity of the HB. This measurement in turn can be used to find the location of those assemblies that are too far away to allow the dwarfs to be seen. From this information, we derive the mean cluster luminosities (absolute magnitudes), which lets us find the distances of those groups that are so distant that the individual stars cannot be seen or analyzed and for which HR diagrams cannot be constructed at all. The resulting spatial distribution of globular clusters allows us to determine the size of our Galaxy and to estimate the distance to its center.

Globular clusters abound in other galaxies. If we assume that they are the same kinds of assemblies that we find in our Milky Way (perhaps debatable; those in the Magellanic Clouds are rather different), we can use the luminosities found by the above procedure to derive the distances of the galaxies in which they reside. This method is especially important for the study of galaxies that are too far away to resolve into individual stars, like M104 in Figure 8.12.

8.9 White dwarfs

We are now in the classic region of the white dwarfs (see Sections 5.8 and 7.10). Although we find them from M (where they are reddish) through early O (blue), the first and best known were found here among the white stars, hence their names. An enormous number carry the 'A' designation, although as we will see later it is something of a misnomer, assigned before we really understood the natures of these stars. The original member of this class is 40 Eridani B, an 11th magnitude companion to the

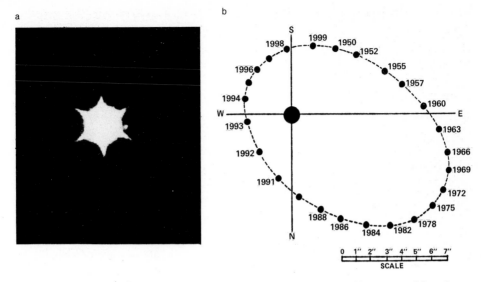

Figure 8.13. (a) Sirius B in the glare of Sirius A, photographed at the Sproul Observatory of Swarthmore College. (b) shows the orbit of the white dwarf about the primary, taken from *Burnham's Celestial Handbook*, Dover Publications, New York, 1978.

brighter K1 V component, which is barely visible to the unaided eye at magnitude 6.0. The most famous, as we know, is Sirius B (Figure 8.13), a real challenge for telescopic viewing because of its proximity to its enormously brighter companion.

We do not employ the usual spectral classes and subtypes when we deal with the white dwarfs. The overwhelming atmospheric pressures wash out the spectrum (Figure 8.3) and leave only the strongest lines, those of hydrogen, the K line of Ca II, helium, and some others. In addition, the chemical compositions are wildly deviant from the stars' main sequence cousins. Consequently, categorization and placement on the HR diagram by traditional spectral classes is misleading if not impossible altogether. The convention in this book is to assign spectral classes according to effective temperatures and to where the stars would fall were they on the main sequence. Even this procedure is uncertain, since without easily measurable line strengths, even the temperatures can be hard to find. We will discuss the problem of classification more fully in the next chapter, after we have completed the nearly full range of their temperatures, and can properly introduce all the varieties. At this point, however, it is appropriate to examine their construction.

One might get the impression here that most white dwarfs are but binary companions. Some are, consistent with the general nature of stellar duplicity, and we concentrate on them, for it is from their orbital motions that we can learn their masses (Section 1.6). But dim white dwarfs are hard to find in the bright glare of main sequence or giant stars, and consequently we know of a much larger number, many hundreds, that are of the single, or field variety. Considering that we cannot see them very far away because of their faintness, they must be very numerous in the Galaxy. Except for the stars that inhabit the lower (K and M) main sequence, there are known to be more white dwarfs than any other kind of star. From a physical point of view, this fact is no surprise, since the class represents the final stage of evolution for all but the most massive of stars.

The strangeness of the white dwarfs is beyond our experience: huge densities, crushing pressures, and occasional magnetic fields that are measured not in thousands but in *millions* of gauss as if they too had been condensed along with the star (recall that the field of the Sun is only about one or two gauss; that of Earth is about one-half). How is this enormous compression possible? We must remember (Section 2.1) that the atom, and matter in general, is mostly empty space: the radius of the hydrogen atom in its ground state is about 10^{-8} cm, whereas the diameter of the proton, the nucleus, is only 10^{-13} cm, so that there is great squeezing room. White dwarfs, in fact, are in but a modest density state compared with their neutron star (pulsar) cousins. But that is a matter for a later chapter.

In our everyday lives a gas behaves according to the *perfect gas law*, wherein pressure times volume is proportional to temperature. There are no restrictions on the energies that atoms or molecules may have, and their motions are describable by ordinary classical Newtonian mechanics. But at white dwarf densities, the free electrons come so close together that quantum mechanical effects become dominant, and the energies they are allowed to assume are highly restricted. In a given volume at a given speed there are a certain number of 'quantum spaces', a sort of volume unit, available. The basis of this science requires that there can be a maximum of only two electrons per quantum space. Any new electrons must be added in unfilled quantum spaces, only at higher speeds. The gas now behaves something like a metal and we say that the electrons are *degenerate* a term that is often applied to the white dwarfs in general (hence the designation 'D').

A main sequence star is supported – kept from collapsing – by its internal gas pressure and by the outward pressure of the intense radiation pouring from its nuclear-burning core. A white dwarf is structured very differently. Although there may still be an ember of fusion left in a shell outside the core, the fierce nuclear furnace of youth has gone dead, the star's feeble glow emanating mostly from heat stored long ago. Its physical support now comes almost solely from the pressure of the degenerate electrons, which by quantum rules can get no closer together. But even these laws have their bounds. If the mass of the star could be raised high enough, even these electrons could not resist the powerful inward pull of gravity. The body would suffer a tremendous collapse, until it was stopped by the pressure of degenerate *neutrons*, created by the mergers of protons with electrons. We would now see a *neutron star*, perhaps only 10 kilometers in diameter, with its matter near nuclear density.

The *white dwarf limit*, first calculated by S. Chandrasekhar, is 1.4 solar masses. No heavier white dwarf can exist. What happens to stars with masses initially larger than this value? They either explode, or as is much more common, lose sufficient matter to get under the limit; recall the enormous quantities of mass lost by the M giants.

But these last paragraphs take us into the whole subject of stellar evolution, a topic that we will explore in depth at a later time. We have yet to look at the truly hot stars of classes B and O, and to examine the fascinating variety of those that do not fit into the standard sequence.

9 *The B stars: beacons of the skies*

Try to imagine the sky without the B stars. We would lose seven of the first magnitude, including two that mark the otherwise rather drab northern hemisphere spring heavens, Spica and Regulus. The Southern Cross would disappear, with the departures of both α and β. The River would have no end, with Achernar's dismissal. Scorpio (Figures 4.1 and 9.1) would be devastated, and unrecognizable (Figure 9.2). The classic seven-star figure of the Hunter would be reduced to Betelgeuse, and δ and ζ in the belt. And if we were also to remove the O stars with which type B is frequently coupled, the latter two would disappear as well, along with his sword, and all but one star (ϱ^2) of his head.

None of the hundred nearest stars has a B spectrum. But among the hundred brightest, they constitute one-third, more than any other, with K giants and A stars well behind at 20% apiece. In this class, we reach an effective combination of number and intrinsic brightness, so that they seem to dominate the naked-eye sky all out of proportion to their true population.

Certainly one of the more famous of the B stars is the formerly obscure Sanduleak (Sk) $-69°\,202$, which blew up as Supernova 1987A in the Large Magellanic Cloud (look ahead to Figure 12.18). This star *used to be* an apparently well-behaved 12th magnitude B3 Iab supergiant, which brightened to third magnitude as a result of the detonation. We will postpone any further discussion of it until Chapter 12.

It is quite common to discuss the B stars in conjunction with their much rarer cousins of type O with which they share many properties: both are hot and blue and produce copious amounts of ultraviolet radiation; in both, the hydrogen lines weaken with increasing temperature; both have some form of helium line present; and they are found in spatial proximity. In certain instances we see the two linked together as 'O and B' or even 'OB' stars.

Even a cursory examination of the sky shows that the O and B stars stick closely to the Milky Way, that broad encircling band created by the combined light of the billions of stars that populate the disk of our Galaxy (Section 1.8; see Figures 1.7, 9.1, and 9.3). Certainly there are prominent exceptions: Regulus, Spica, η Ursae Majoris (Alkaid), and some others appear well away from this pale ribbon, but only because they are relatively nearby, and just happen to appear 'overhead' of the Sun relative to the rest of the disk. When we plot the galactic positions of the O and B stars in three dimensions, we discover that they actually all lie within about 100 parsecs of the centerline of the

Figure 9.1. Our Milky Way, where we find the O and B stars, seen here in Ophiuchus and Scorpio. The lower portion of the latter constellation is in the lower right quadrant. The head of Scorpio is shown in Figure 4.1. Antares is at far right center, and again appears relatively dim because of its red color. The structure within the Milky Way is caused by absorbing clouds of interstellar dust. From *An Atlas of the Milky Way*, by F. Ross and M. Calvert, University of Chicago Press, Chicago, 1934.

galactic plane. The ring of the brighter hot stars around us is known as the 'Gould Belt', after the nineteenth century astronomer B. A. Gould. It is actually tilted slightly off the Milky Way, a result of a local distortion of the disk of the Galaxy.

This relationship between the hot stars and the galactic plane is only part of a broad correlation between spectral type and galactic distribution; at the other extreme, the M stars show a lesser confinement to the disk, and can scatter to great distances into the spheroidal galactic halo. This distinction is the result of different stellar ages. Because the massive and brilliant O and B stars burn their nuclear fuel at such furious rates and do not live very long, they must be found near their birthplaces, where we also find the great concentrations of the interstellar gas and dust out of which they were

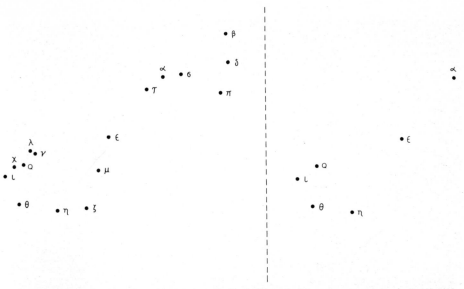

Figure 9.2. Scorpio with and without its B stars.

Figure 9.3. NGC 4565 a spiral galaxy seen edge on. The O and B stars are confined to the thin disk, as they are in our own Galaxy. The dark line down the center is caused by internal interstellar dust. The same phenomenon can be seen in M104 in Figure 8.12 and in our own Milky Way (Figure 1.7). National Optical Astronomy Observatories (Kitt Peak) photograph.

Figure 9.4. The Pleiades. The inset shows the whole cluster, the larger one a magnified view of the central region. All the brighter stars are class B, led by Merope (η Tauri, B5 III), the left-most one in the main photograph. The delicate wisps that surround several of the stars are reflection nebulae. Large photo from National Optical Astronomy Observatories (Kitt peak); inset, Lowell Observatory photograph.

created, namely in the plane of the Milky Way. Many of the M stars, on the other hand, are very old; they were born in the halo long before the disk developed, and their positions also reflect their places of origin since there are no forces that could compress them into the disk.

Even within the galactic circle, we find a strong predisposition of clumping among the hottest stars, as the two great mythological enemies Scorpio and Orion bear witness. These loose groupings are known as *O* or *OB associations*, and we will look at them more closely when we review the O stars in the next chapter. The most luminous B star of all is known only as 'Cyg OB2 No. 12', the 12th star listed in the number two OB association in the constellation Cygnus.

We cannot leave this discussion without a glance at the most famous of all clusters, the Pleiades of Taurus, shown in Figure 9.4. *All* of the classic Seven Sisters are B stars, as are several other brighter members. Recall from Section 5.7 that this cluster is instrumental in the calibration of the HR diagram through the technique of main

Figure 9.5. The HR diagram for classes B and A. The main sequence rises rapidly to high luminosities. The Pleiades stars, typified by Merope fall here. Sk −69° 202 is the 12th magnitude B star in the Large Magellanic Cloud that exploded to produce Supernova 1987A. See Figures 3.8 and 12.7 for credits.

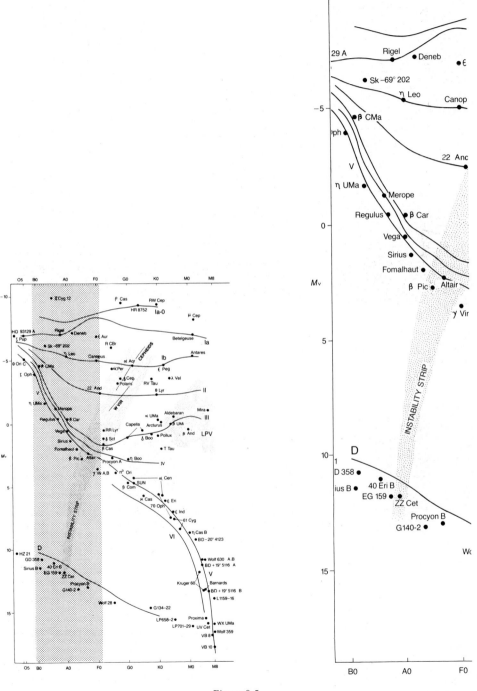

Figure 9.5.

sequence fitting. Without this gem of a group, whose evening visibility ushers in the northern winters, whose various apparitions have signaled both planting and harvest times, and which is known by all cultures, we would be very much the poorer. Figure 9.5 displays the array of the B stars, including the Pleiades' Merope, as we begin to approach the top of the main sequence.

9.1 Classification

As we pass 9 500 K on our way along the main sequence into the B stars, the hydrogen lines at last begin to fade (Figure 9.6), as indicated by the assignment to this group of the second letter in the alphabet (see Section 3.3). At this critical temperature, we achieve a maximum relative number of hydrogen atoms with second orbit electrons, from which the Balmer lines arise. Above 9 500 K, H begins rapidly to ionize, and even though the ratio of the number of second orbit electrons to that in the first continues to increase, the total quantity of Balmer absorbers goes down along with the count of neutral atoms.

At the same time, we see in these spectra the rise of the neutral helium lines, which make their first appearance at about B9. These absorptions arise also from the second orbit, but for helium the corresponding energy level is much farther from the ground state than it is for hydrogen, and a considerably higher temperature is required in order

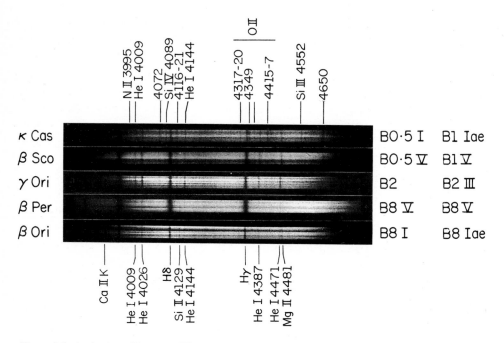

Figure 9.6. A selection of B spectra. The strengths of the hydrogen lines decrease toward earlier subclasses as the H becomes ionized, and the He I lines increase to a peak around B2. Positive and negative luminosity effects for Si III and H respectively are easily seen. From *An Atlas of Stellar Spectra* by W. W. Morgan, P. C. Keenan, and E. Kellman, University of Chicago Press, Chicago, 1943.

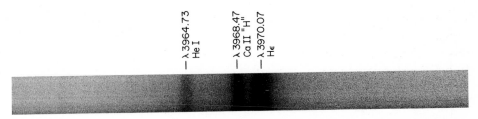

Figure 9.7. A very high dispersion segment of the spectrum of Rigel, β Orionis, a B8 Iae supergiant. In this 37 Å span (λ3951 Å to λ3987 Å) only three lines can be seen, one each of hydrogen and helium, and the 'H' line of Ca II. In the spectra of later type stars Fraunhofer H is inextricably blended with Hε. Here the lines are weak enough that we can see them resolved (compare with the several lower dispersion spectra of later types presented before). The Ca II lines have been weakening since late G, and the hydrogen absorptions are fading from their maxima at early A. The myriad metal lines of low ionization, so evident in previous high dispersion spectra, have vanished, victims of the higher temperature. Mt. Wilson and Las Campanas Observatories, Carnegie Institution of Washington, photograph courtesy of R and R Griffin.

to pump electrons into it. He I strengthens up to B2, and then with the rapid onset of helium ionization begins to fade. This spectrum is then replaced by the lines of He II that make their appearance at B0 and develop on into O. Most metallic lines are gone or are quite weak (Figure 9.7), but we find a number of absorptions of higher ionization states of silicon, oxygen, and carbon. Numerous Draper classification criteria are available, for example, the ratios of Si IV or C III to He I in early B (which increase with temperature), and Si II or Mg II to He I in late B (which decrease).

The B stars present us with a problem in classification because the type is so broad. They encompass an enormous temperature range, from 10 000 K at B9 to nearly 30 000 K at B0, which reflects a similar large dispersion in masses between roughly 3 and 20 times solar. Here, the ordinary decimal subtypes are insufficient; we must further subdivide, with B0.5 and B9.5 becoming standard. The difficulty is particularly apparent when we examine the various luminosity types. With some overlap, different rules are used for giant and supergiant temperature sequences, and each Draper subclass has its own set of luminosity criteria.

The common theme among all spectral subclasses is a negative absolute magnitude effect for the hydrogen lines (see Section 8.1); the Balmer absorptions are considerably weaker among the supergiants, and show little change with subtype. A parallel effect in He II is so large that the lines are easily noticeable in B0 dwarfs, but are gone in the B0 supergiants. A sample of other luminosity criteria include the line strength ratios of Si III and Si IV to He I at B0, O II to He I at B1 and B2 (see Figure 9.6) and C II to He I at B3 and B5, all of which increase with dropping absolute magnitude.

The class B stars are very prominent in the ultraviolet. As we proceed upward along the main sequence, progressively more radiation spills into the that region of the spectrum shortward of the transparency limit imposed by the Earth's atmosphere at a wavelength of 3000 Å (Figure 1.10). G and K stars emit only a small amount of light below about λ2600 Å; by late F to mid A the stellar cutoff has moved shortward to

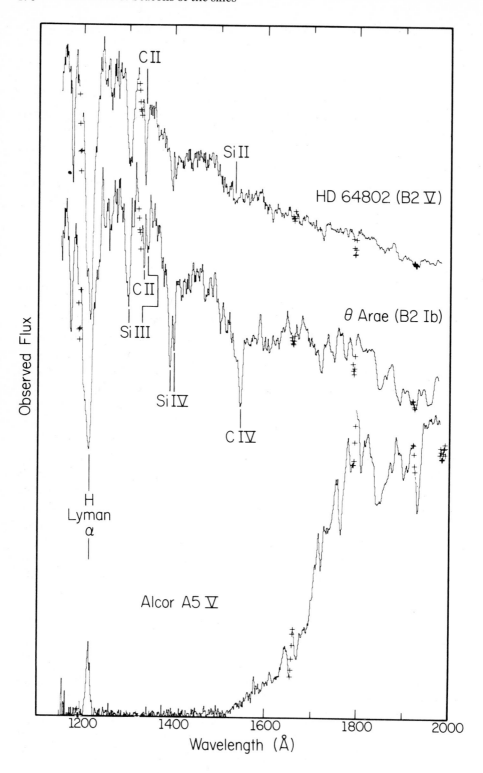

Figure 9.8.

roughly $\lambda1600$ Å. In early A, in response to increasing temperature, we begin to fill in nearly all the accessible ultraviolet down to Lyman α at $\lambda1216$ Å, and in type B, the UV actually dominates, as we see in Figure 9.8.

The ultraviolet has many line features that are sensitive to MKK class. In particular, the Si IV and C IV lines increase from dwarf to supergiant and the C II strength drops. We also observe that species of lower ionization decrease in strength as we increase the temperature along the main sequence from A5 to B2, resulting in a simplification of the spectrum. The ultraviolet has only been opened to observation relatively recently with the advent of orbiting telescopes, and has become profoundly important to our understanding of the natures of the stellar atmospheres, including temperatures, compositions, and as we will see in the next two chapters, the powerful winds that blow from their surfaces.

9.2 More strange dwarfs

Just as in type A, we see several categories of B stars with strange spectra. The magnetic stars that we examined in Section 8.3 extend well into this class to roughly B5, where they are called Bp. There are other interesting kinds of non-magnetic stars. Between about B9 and B4 lie the *mercury* or *mercury–manganese* (*Hg–Mn*) stars, with enormous overabundances of these elements, as well as of such others as phosphorus, gallium, and yttrium. Other elements: aluminum, calcium, and nickel, are oddly underabundant. The helium lines present us with further anomalies. We see stars in which they are either abnormally strong or unusually weak. Presumably, at least some of these effects are due to some sort of elemental diffusion and separation (Sections 7.4 and 8.5).

9.3 Be and shell stars

The average stellar rotation speed reaches a maximum in class B. That is not to say that slow rotators do not exist within the set; it is among these, as before, that we find the chemically peculiar stars cited above. But at the other end of the scale, the velocities can become quite enormous, well over 200 km/s. The large spin rate seems somehow to be related to the ejection of matter that forms an equatorial ring around the star that radiates emission lines. These Be stars have highly characteristic hydrogen emission features superimposed upon broad, strong absorptions produced in the stellar

---

Figure 9.8. The ultraviolet spectra of three stars taken with the short-wavelength spectrograph camera of the *International Ultraviolet Explorer* (*IUE*) satellite. Since the data are acquired by digital counting techniques, the spectra are necessarily presented in graphical form. At the top are two B2 stars with temperatures near 20 000 K: the 5th magnitude dwarf HD 64802 in Puppis, and the 4th magnitude supergiant θ Arae. The enhancement of higher ionization toward higher luminosity is seen by the growth of Si IV and C IV and the diminishment of C II. Below is the spectrum of the A5 dwarf Alcor. The lower temperature of only 8100 K produces the sharp dropoff in the ultraviolet intensity near $\lambda1500$ Å. The numerous lines of lower ionization in Alcor's spectrum are absent in that of the higher temperature dwarf. Note also the strong hydrogen Lyman α (the level 1 to 2 transition) absorption apparent in the B stars. The emission feature in Alcor's spectrum comes not from the star but is hydrogen Lyman α produced in the neighborhood of the Earth. The '+' marks indicate breaks in the data. From the *IUE Ultraviolet Spectral Atlas, International Ultraviolet Explorer NASA Newsletter No. 22*, 1983.

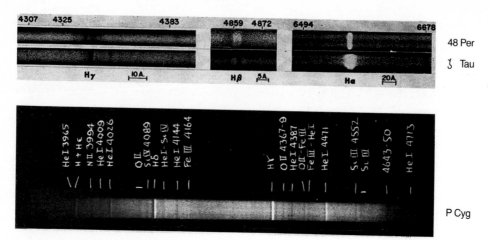

Figure 9.9. Three emission line stars. The upper two strips show sections of the spectra of 48 Per (B3 Ve) and ζ Tau (B4 IIIpe). The latter is a shell star. Note the doubling of the central emission feature, indicative of an orbiting circumstellar ring. The lower strip is a spectrum of P Cygni (B1 Ia). Both the H and He I lines show violet-displaced absorptions characteristic of an outflowing wind (see Figure 9.10). Upper: from an article by A. B. Underhill in *Stellar Atmospheres*, J. L. Greenstein, ed., University of Chicago Press; lower: from *An Atlas of Stellar Spectra*, by W. W. Morgan, P. C. Keenan and E. Kellman, University of Chicago Press, Chicago, 1943.

atmospheres (Figure 9.9). The emission caused by the orbiting circumstellar material is broadened by Doppler shifts, as one side of the ring will be coming at us, with the other going away. At the very center of the system the gas moves across the line of sight where it appears more opaque and absorbs some of its own radiation, creating a central reversal, or absorption. The emission line thus looks doubled, with a blue approaching component and a red receding one. The ratio of their strengths varies with the amount and distribution of mass, which can be quite irregular. Sometimes the ring and its emissions disappear altogether, to reappear some years later.

The rings come in a range of thicknesses, and at the high end of the scale we find a subset, the *shell stars*. Here, the amount of mass is sufficient to superimpose metallic absorption lines onto the background spectrum. Many Be stars are familiar to the casual observer. The best known are γ Cas (B0 IVe), ζ Tau (B4 IIIpe), the Pleiades' Pleione (28 Tau, B8 Vpe), φ Per (B2 Vep), and 48 Lib (B5 IIIpe), all of which are shell stars.

In spite of decades of study we still do not know what triggers the ejection, and why some rapid rotators develop this feature, whereas others do not. One idea suggests that a stellar wind, driven outward by the star's radiation, is enhanced and modified by the rotation and associated magnetic fields. A newer notion links the phenomenon with a strange form of stellar variability. The common variables, the Cepheids and the Miras for example, are said to pulsate *radially*, that is they expand and contract inwardly and outwardly along any radius of the sphere, with no dependence on stellar latitude or longitude: the stars simply get bigger, then smaller. However, we have discovered that stars can also pulsate *non-radially*, in a complex manner that depends on longitude or upon both that and latitude. If we could watch such a star, one part would be moving outward and one part inward, something like a blob of oscillating gelatin.

Although these 'jiggles' may produce only a subtle overall magnitude variation, and are detectable mostly by changes in the shapes of absorption lines, they appear to be powerful enough to be able to drive the winds of the later-type B stars where the radiation mechanism does not work well. These oscillations now appear to be very common on the upper (O and B) main sequence. Where coupled to rapid rotation, they could produce the Be stars.

9.4 Stellar stability

This is a good point at which to comment on the stability of the main sequence and of stars in general. We traditionally divide the stars into the variables and non-variables. But does the latter group exist at all? As our measurement techniques get better and better, we find that more and more stars vary, and at some level, however subtly, they probably all do, even if only by a few thousandths or ten-thousandths of a magnitude. On the upper main sequence we see the above non-radial oscillations; in class A we have the effects of the Cepheid instability strip; and then further down, where dynamo action and chromospheric activity set in, we have star spots and flares. The sky appears so steady fixed and stable as we look at it – but only if we do not look too hard.

9.5 Mass loss and P Cygni lines

Amongst the B stars we also begin to see a number whose spectra exhibit what we call *P Cygni features* (see the lowest spectrum in Figure 9.9). These are emission lines that are flanked at their violet edges by absorptions. They are characteristic not of a circulating, circumstellar ring, but of an expanding envelope, or stellar wind. The Doppler-displaced absorption is created by opaque gas that is flowing outward in front of the star directly toward the observer, as diagrammed in Figure 9.10. P Cyg itself is a fifth (apparent) magnitude B1 Ia supergiant, whose Roman letter name derives from a largely defunct system of nomenclature (see Section 1.2). It does not really define a class of stars, however, as its distinctive spectral nature is shared by a variety of types from early A (for example, Deneb, which exhibits a weak P Cyg line profile at Hα) through O, and beyond into the domain of the planetary nebula nuclei.

9.6 Giants

The B giants, of luminosity class III, lie only about one magnitude above the main sequence, and are merely about 60% larger than their so-called dwarf counterparts, which themselves are two to three times the solar diameter. Among the earlier subclasses, we encounter an interesting set of variables known as β Canis Majoris (Murzim, once again), or β Cephei stars. This group constitutes somewhat under 10% of the B1–B2 subgiants and giants. The periods are measured in hours, and the amplitudes are but a tenth of a magnitude or so. About half exhibit two distinct periods of variability that beat against one another to cause a third period (much like the slow oscillation heard when two not-quite-similar audio tones are played on musical

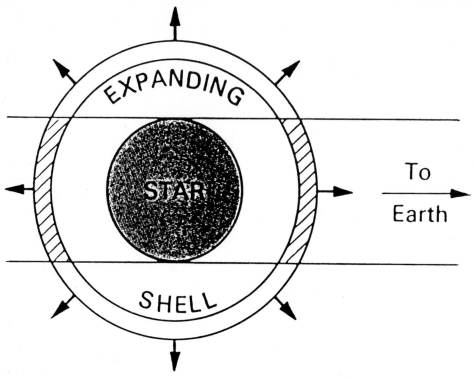

Figure 9.10. The formation of a P Cygni line. The outflowing gas not seen projected against the star produces an emission line. Since most of the motion is perpendicular to the line of sight it will not be Doppler shifted, but will appear to be broadened. The gas that lies in front of the star, however, must produce an absorption feature in the stellar continuum. Since this part of the wind is coming straight at us, the absorption will be Doppler shifted to shorter wavelengths and will appear at the violet edge of the emission line. Outflow velocities can be found by the degree of shift. From *Principles of Astronomy, A Short Version*, 2nd edn., by S. P. Wyatt and J. B. Kaler, Allyn and Bacon, Boston, 1981.

instruments.) Radial velocity studies show that, like the Cepheids, the stars are pulsators, except that maximum luminosity coincides with the smallest radius rather than with greatest expansion velocity (Section 7.5). They are generally slow rotators and seem to be in a particular state of internal evolution.

The enormously luminous B supergiants are considerably smaller than their class M counterparts. Whereas μ Cephei's (M2 Iae) radius is about 11 AU, that of Cyg OB2 No. 12 (B5 Ia) is 'only' one AU. The difference, of course, is due to the much higher temperature of the early type star, which radiates roughly 300 times as much energy per unit surface area.

9.7 Dust and nebulae

The proximity of the young O and B stars to their parent interstellar clouds allows this matter in turn to be illuminated. The plane of the Milky Way is therefore filled with bright *nebulae*, some of which, like the great Orion Nebula, are favorite targets for the backyard telescope.

The B stars are responsible for a specific type called the *reflection nebula*. Everywhere that we find free gas between the stars we also find *dust*. This term is a loose expression for fine solid grains of material typically a ten-thousandth to a thousandth of a millimeter across. Roughly one percent of the mass of the interstellar matter is in the form of these particles, which appear to be made largely of carbon or of silicon and oxygen in the form of silicates. The origin of this material and the processes that produce it are still unclear. As we saw earlier, it is produced in ample quantities in the expanding outer atmospheres of M giants, Miras and carbon stars. Differing stellar compositions result in different kinds of grains, which are then further modified by their long exposures in the cold depths of space. The Pleiades, pictured in Figure 9.4, provide a marvelous example of a reflection nebula. They are embedded inside a concentration of interstellar matter. The grains within the delicate wisps of gas reflect, or more accurately, scatter the light of the embedded B stars.

As we move toward earlier subtypes, a quite marvelous transformation occurs at B1. At an effective temperature of about 25 000 K, the star begins to radiate a significant fraction of its light shortward of the critical Lyman limit at a wavelength of 912 Å. Within this spectral region the energy of an emerging stellar photon is sufficient to ionize a hydrogen atom (Section 2.7). The star's ultraviolet light then ionizes the nearby gas out to a distance at which all the ultraviolet radiation is used up. The recombination of the electrons and protons causes this gas to glow, and we now see a *diffuse nebula*, or *H II region*, or *Strömgren sphere* (since in a uniform gas the nebula will appear as a spherical bubble). These, like the Orion Nebula, are most prominent when they are associated with the more luminous O stars, and we will discuss them more fully in the next chapter.

The two types of nebulae can easily be discerned by their spectra: the reflection nebulae show the absorption lines of the nearby star scattered off the dust, and the diffuse objects exhibit a bright, or emission-line spectrum caused in part by the recombining, downward cascading electrons. An easier test is simply color. The reflection nebula will appear blue, the same color as the star. The H II region, however, is decidedly red, the hue being produced by the strongest hydrogen emission line, Hα, at $\lambda6563$ Å. The two can easily be discriminated in color photographs, or simply by comparing black and white images made in the blue and red spectral regions.

The dust that makes the reflection nebulae also inhabits the H II regions, giving them a small reflection component. More importantly, it pervades the entire disk of the galaxy. Without a source of illumination, the dust will appear dark against the stellar background, as its scattering properties allow it to dim the light of distant stars (Figure 9.11). These tiny grains cause much of the structure of the naked-eye Milky Way that is so dramatically evident in Figure 1.7, such as the prominent rift that begins in Cygnus and extends to Scorpio. Because of the absorption, external galaxies are seldom seen in the direction of the plane of our own Galaxy. Blank areas of the sky that appear as gaps in the splash of the Milky Way, like the Coalsack near the Southern Cross, are actually nearby dense absorbing clouds unlit by embedded stars. They are so prominent in the southern hemisphere that the Incas of Peru and Chile actually made constellation

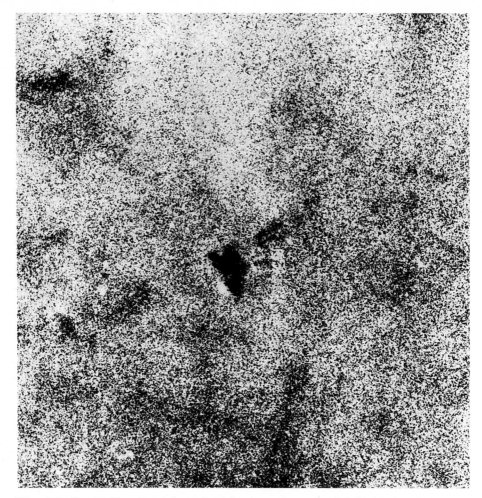

Figure 9.11. Barnard 86, an interstellar, or Bok (after the astronomer Bart J. Bok) globule, a thick cloud of dust that blocks the light of background stars. Other, lesser, dark structures pervade the photograph. ©1960 National Geographic – Palomar Observatory Sky Survey, reproduced by permission of the California Institute of Technology.

figures of them. The cloud temperatures are so low that the grains can become coated with ices. It is in the most compact of them that new suns are beginning their lives.

9.8 Dust and interstellar reddening

The dust produces an immensely serious problem in the calculation of stellar distances. We saw much earlier that we might determine how far the stars are if we could infer their absolute magnitudes from their spectra, or in the case of Cepheids, from their periods. To do so, we employ the magnitude equation, $M = m + 5 - 5 \log D$ (Section 1.13), where D (in parsecs) is the unknown quantity. However, if there is absorbing dust in the line of sight, a star will appear fainter than it should, m will be too large, and we will overestimate D. One magnitude of dimming, or *extinction* will make it look 60% too far away, and several magnitudes is not uncommon.

Fortunately, the dust does not absorb uniformly across the spectrum: it is about twice as efficient at short optical wavelengths as it is at long. Thus the grains not only dim starlight, they *redden* it. Normally, a mid-B star would have a color index (the difference between the star's photographic and visual magnitudes, or $B-V$: see Section 1.14) of about -0.2 and would appear blue (Figure 3.10). But it is not uncommon to find one looking as yellow as a G star (color index $+1.2$), or on occasion, even redder (2.0 or so). By comparing dust-reddened and unreddened stars of the same spectral class, we can determine the relation between the visual absorption in magnitudes, A, and the degree of reddening. The latter is referred to as the *color excess*, $E(B-V)$, the difference between the observed color, expressed in magnitudes, and that which would be observed were there no dust in the way. The ratio A/E is generally found to be equal to three. Once E is known, A can be found, subtracted from m, and a more realistic value of D derived.

The color excess can be measured by determining the star's intrinsic (un-reddened) color from its spectral class and the correlations plotted in Figure 3.10, and subtracting it from the observed value. We can also employ colors alone from the UBV system of photometry (Section 1.14). For any given luminosity class there is a strict relationship between the colors $U-B$ and $B-V$. Those for the main sequence and for supergiants were discussed in Section 3.7 and displayed in Figure 3.11. As we increase the reddening, for a specific kind of star, however, the color excesses $E(B-V)$ and $E(U-B)$ change in relation to one another in a considerably different fashion. For example, the B2 dwarf in Figure 3.11 will change its location on the plot along the locus labeled 'R'; loci for other stars will be parallel to it. We need then only locate a star on the color–color plot and trace it up and to the left along a reddening locus until it meets the standard curve, which gives its intrinsic color and spectral class simultaneously with the reddening.

For a single star we would have to know or assume the luminosity class since the standard curves are different. In addition, the hook in the main sequence curve in the neighborhood of the A stars caused by the Balmer discontinuity can produce an ambiguity: a reddened B8 star could look either like an unreddened F0 or a K0. We can resolve these problems by using the array of stars in clusters, for which this method is eminently suited, since we can identify an extensive main sequence.

However, after all this labor nature can be unkind. In some peculiar regions of space, the quality of the grains can be different and A/E can be larger than 3, still leaving us with some uncertainty.

9.9 Polarization of starlight

In addition to reddening starlight, the dust grains also *polarize* it. We encountered this phenomenon briefly in Section 8.4 in conjunction with magnetic stars and the Zeeman effect. Now let us go into the matter a bit more deeply. In normal radiation, the individual waves, or photons, oscillate in random directions, that is there will be just as many waves vibrating in one plane as there will in any other. But in polarized light, there will be a preferential direction: there may be more moving 'up and down' than 'sideways'. If the polarization is *perfect*. all the waves will oscillate in the same direction, much as we would draw them in the plane of a sheet of paper. We characterize the degree of the phenomenon by a percentage figure.

Light (and radiation in other parts of the electromagnetic spectrum) is commonly polarized in nature. Sunlight reflected from a rough surface is partially polarized, as is the radiation from the blue sky (which is caused by the scattering of sunlight from air molecules). We can polarize light by passing it through a filter in which the crystal structure has a specific alignment, like a comb: radiation vibrating parallel to the teeth gets through, whereas that oscillating perpendicular to them is blocked. This principle is used in the construction of polarizing sunglasses. The crystal structure is aligned relative to the direction of polarization inherent in the reflection of sunlight from a roadway, which blocks the glare, rendering driving much more comfortable. By rotating the lens from a pair of these sunglasses, one can also easily observe the degree and direction of the polarization of skylight.

The interstellar grains are irregular in shape, and are at least in part brought into alignment by the weak (10^{-6} gauss) but pervasive magnetic field that inhabits the galactic disk. Thus they act like a polarizing filter, and as the light is reddened it is also partially polarized. The degree of this phenomenon can rather easily be measured by examining stars with a rotating polarizing filter attached to the telescope. We can then use these data to explore the galactic magnetic field.

9.10 White dwarfs

We return again to the white dwarfs as they stretch across the lower end of the HR diagram. Now, with most of the stars in place we can consolidate our earlier discussions (Sections 5.8, 7.10, and 8.9) and conclude the matter. These stars are denoted by the letter 'D' (for degenerate), and in this series have been assigned a location on the diagram according to their effective temperatures and the main sequence temperature scale. Why is this odd procedure necessary? Even though the spectra of degenerate stars do not look much like those on the main sequence, they were indeed originally assigned the standard categories based upon whatever lines were present. Thus white dwarfs with strong hydrogen lines were called DA, and those with He I, DB. Other classes – DF, DG, DK, and DM – were allocated according to the strength of Ca II either alone or relative to the strengths of other lines; see Figure 9.12 and Table 9.1.

But something very peculiar is happening here. When we carefully examine a typical DB star we see no trace of hydrogen lines, quite in contrast to an ordinary B dwarf, where they are still quite strong. And when we analyze a representative DA, we see prominent hydrogen lines, but little or nothing else. In fact it has become quite apparent that the classical designations, when applied to white dwarfs, have less to do with temperature and far more to do with chemical composition.

In more modern times, the white dwarfs have been chemically separated into two very broad primary groups whose names have been derived from spectroscopic

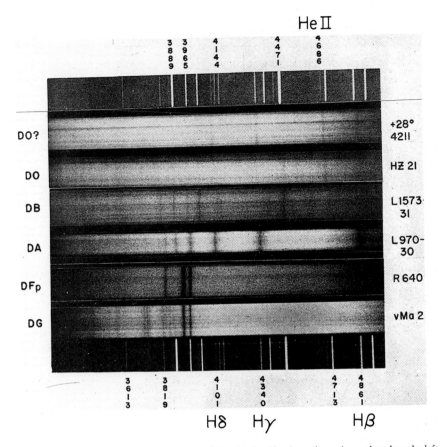

Figure 9.12. A sequence of white dwarf spectra. The old classification scheme is employed on the left, the star names are on the right. All lines without identification in the top three spectra are He I. Note the absence of hydrogen lines from the DB spectrum, which displays only helium. The two broad lines in the DF and DG spectra are H and K of Ca II. The new-style classifications (Table 9.1), from HZ 21 down, are DO2, DB4, DA4, DZA6, and DZ7, where the number is derived from the temperature as described in the text. Palomar Observatory spectrograms, from an article by J. L. Greenstein, in *Stellar Atmospheres*, J. L. Greenstein, ed., University of Chicago Press, Chicago, 1960.

notation, but which now mean little in the way of traditional classification. The line-forming atmospheres of the DA stars are nearly pure hydrogen, whereas those of type DB seem to be practically all helium. Both kinds span a wide range of temperature that extends over most of the classical spectral types. That is, the DA stars are no longer confined to the neighborhood of 10 000 K as are the class A dwarfs. They are seen well into what would ordinarily be called class O at 50 000 K, down to temperatures at which the H lines become too weak to be seen and the spectrum becomes continuous, where they may be called DC. Some of the cooler white dwarfs, formerly classified as DF through DM on the basis of their metallic spectra (usually calcium or iron), may chemically actually be type DB. The hot ends of both sequences are sometimes called DO. Further complicating the situation, there are several hybrid stars that combine the above characteristics, and some that exhibit spectral features of carbon.

Table 9.1. *White dwarf classifications*

	Old			New	
DC	continuous		DA	H only	
DO	He II strong, with He I or H		DB	He I, no H or metals	
DB	He I strong, no H		DC	continuous	
DA	H present, no He I		DO	He II, with He I or H	
DA, F	H weaker, Ca II weak		DZ	metal lines only, no H or He	
DF	Ca II, Fe I		DQ	carbon present	
DG	Ca II				
DK	weak Ca II				
DM	Ca II; Ca I weak; TiO?				
C_2	carbon molecular bands				
	Suffixes (or subscripts)			Suffixes (final letter)	
p, P	polarization, or peculiar (PEC)		P	magnetic stars with polarized[a] light	
wk	weak		H	magnetic stars without polarized[a] light	
e	emission		X	peculiar or unclassifiable	
s, ss	sharp and very sharp		V	variable (ZZ Ceti, or other)	
n	diffuse				

[a] See Section 9.9

The old system, on the left, follows the order of classical spectral types. The new system uses many of the same letters, but describes only spectral features, and is divorced from color. The second letter denotes the primary characteristic, an optional third any secondary properties, using the same letters, and a fourth any peculiarities. An appended number, based on $50\,400/T$, describes temperature. Adapted from an article by E. Sion, J. Greenstein, J. Landstreet, J. Liebert, H. Shipman, and G. Wegner that appeared in the *Astrophysical Journal*.

What is the origin of these two general sequences, the DA and DB white dwarfs? In the case of the former, it appears to be atmospheric separation of elements heavier than hydrogen. Because of high surface gravities, the line-forming atmospheres of degenerate stars are only a few hundred meters thick. The helium has simply settled to the bottom. In the instance of the DB types, there *is* no hydrogen. The envelope of the star has been completely lost through stellar winds and/or nuclear burning, leaving behind the old spent core that long ago transmuted its hydrogen into helium. In spite of these apparently simple explanations, however, we still do not understand many of the physical processes involved. The matter is further complicated by the possibility that the white dwarf atmospheres may be contaminated with gas accreted from interstellar space or from a binary companion.

With the white dwarfs, we are faced with a truly confusing situation, in which the Draper classes are used to describe both composition and temperature. In order to bring some order to the problem, and to facilitate intelligent discussion of the different types, a new classification system has recently been developed (Table 9.1). A team of white dwarf specialists has suggested a code consisting of two to four letters followed by a number. The first is still 'D'. The second, for familiarity's sake, employs some of the same symbols currently used, but now only to identify the lines that appear. DA is used only when hydrogen is exclusively present, and DB for only helium. DF, DG, and DK are fused into DZ, as 'Z' is a universally understood code for 'metals' (here, any heavy element other than carbon.) The carbon white dwarfs become DQ; C cannot be used because of confusion with 'continuous'.

The third letter is optional, and describes secondary characteristics, with the same code as above. For example, a DB star with a trace of Ca II becomes DBZ. A final letter (that may be third or fourth in order) describes any peculiarities that may exist, even though the description may be beyond the bounds of what is normally considered to be spectral classification: see the table. A DA star with a magnetic field, but whose radiation is polarized by the star's field, not by interstellar grains (see Sections 8.4 and 9.9) becomes DAP; if it has a trace of carbon, it is DAQP.

Finally, we append a number that describes temperature, derived by rounding $50\,400/T$ to the nearest integer. (The parameter $\theta = 5\,040/T$ is commonly used in stellar atmospheric studies.) This suffix is the most controversial, as it cannot be found directly from a simple examination of the spectrum, and is therefore not really a form of spectral classification. The temperature must be found by other means, such as the examination of the shapes of the lines, or by the way in which the continuum intensity changes with wavelength. However since the spectrum does not correlate well with temperature, many astronomers believe the introduction of physical parameters is necessary to the understanding of these stars.

Similar controversy involves the P and H suffixes, which denote the existence of polarization and the presence of magnetic fields. Polarization cannot be seen by eye on a classification spectrum: its detection requires sophisticated instrumentation. Yet it is another important characteristic of these stars, and deserves some sort of notation. The magnetic fields seem to have been compressed along with the stellar volume, and can be amazingly strong. At the top end, near an astounding 300 million gauss (compare with the 34 000 gauss field found for the strongest Ap star), the hydrogen atoms become so distorted that they can produce line features (the so-called *Minkowski bands*, the strongest of which is at $\lambda 4135$) well away from the standard Balmer series positions, thus actually allowing traditional classification. However stars such as these are exceedingly rare, and if we want to include magnetism in the spectral description we must adopt other detection methods. We have to keep in mind, however, that we are then outside of the field of classification as it was originally conceived, which was designed to be *independent* of measured physical parameters.

9.11 White dwarf variables

Among the white dwarfs the temperature range that is characteristic of class B main sequence stars is the final setting for the instability strip. Here we find a few odd variables called *ZZ Ceti* stars, which have multiple periods measured in minutes. They are non-radial pulsators (Section 9.3), and the short oscillation times are, as we would expect, the result of the tiny stellar radii. Although formally in the B spectral range, between roughly 10 500 and 13 000 K (the exact figures are controversial), they are actually all DA stars. The pulsations of the ZZ Ceti stars are consequently driven by the zone below the surface in which hydrogen is in a state of partial ionization (see the discussion of Cepheids in Section 7.9.) A bit farther up the white dwarf sequence, at roughly 20 000 K and still within the temperature range of the B stars, we find another zone of instability where the DB stars are allowed to oscillate with pulsations driven by helium ionization. These are quite rare, however.

We see from this examination of A stars and white dwarfs that the old science of spectral classification, however it is defined, is still very much alive. The systems must evolve so as to bring order to new astrophysical discoveries.

10 Class O: the head of the spectral sequence

And at last, after our climb of seven steps, we arrive at the top to encounter the spectacular stars of type O. What superlatives we may use: the hottest, bluest, brightest, most massive, and rarest! It is this last quality that makes the O stars relatively unfamiliar to the naked eye observer: none are of the first magnitude and we see only four of the second. This quartet does, however, contain two of the most prominent stars in the sky, δ and ζ Orionis, the eastern pair of the Belt. Both are binaries. The first has an O9 main sequence component coupled to a B0 giant companion, and the second is dominated by an O9.5 Ibe supergiant. To the eye they appear similar to the third belt member, ε (B0 Iae).

The other two have early subtypes and are well known to southern hemisphere observers. Both ζ Puppis, an O5 Iaf supergiant, and γ^2 Velorum, which contains an O7.5e component, add life to the sparkle of the southern Milky Way. The latter is especially remarkable: the O7.5 star has a brilliant unresolved *Wolf–Rayet* O-type companion (see Section 3.4; we will look closely at these odd emission-line stars below.) There are four other visual companions associated with it, including γ^1 Velorum, a fourth magnitude B1 subgiant. Altogether, the system is a favorite sight for an observer with a small telescope. These two stars, ζ Pup and γ Vel, lie within an extraordinary volume of space outlined by the great Gum (after Colin Gum) Nebula, the vast remnant of an exploded star. Northerners are less privileged: to find a fairly early O above the equator they must look to magnitude four, to ξ Persei (O7e).

The perception of this earliest stellar type by the telescopic observer is quite different. The O stars energize the gaseous nebulae that constitute a prominent part of the repertoire of the backyard astronomer. The best known of any of this class is certainly θ^1 Ori C, the brightest of the Trapezium, the O6 powerhouse that lights the Orion Nebula (Figure 10.1). As we go deeper into space we encounter many others that cause the Milky Way (Figures 1.1, 4.1, 9.1) to be decorated with these beautiful objects. For the devotee of galaxies, the young O stars and their associated clouds of birthing gas dominate the spiral arms (Figure 10.12 below) and provide the tracers for active regions of star formation.

The most luminous of this type known with certainty is HD 93129A, an extraordinary seventh magnitude O3 If star in the great Carina nebula near the odd variable η Carinae (it appears in Figure 11.12, and its remarkable ultraviolet spectrum

203

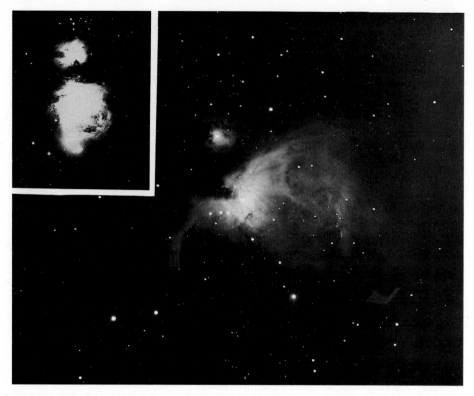

Figure 10.1. The great Orion Nebula. The main view shows the delicate inner structure. The famous Trapezium, θ^1 Ori, is at the center of the brightest portion; the O6 C-component is the dominant ionizing source. The odd-looking feature at the lower right is an internal reflection in the telescope. The $\frac{1}{6}$-scale inset at the upper left displays the full extent of the object, which is some 7 parsecs across. The upper nebula in the wide angle photo is NGC 1977, which is out of the field of view of the other picture. The nebula is set within a much more extensive gaseous complex, and its context can be seen in the full photo of Orion in Figure 1.1. Center: University of Illinois Prairie Observatory photograph; inset: ©1960 National Geographic–Palomar Sky Survey, reproduced by permission of the California Institute of Technology.

is shown in Figure 10.5 below). Note on the HR diagram in Figure 10.2 that this star seems less luminous than Cyg OB2 No. 12, our brightest B star. We again encounter, as we did for type M, the matter of the bolometric correction (Section 4.3). At the lowest stellar temperatures most of the energy lies in the infrared, so that the stars are actually brighter than indicated by their absolute visual magnitudes. Corrections are some −3 magnitudes at M5. At the hot end of the stellar range, most of the radiation falls in the invisible ultraviolet and, again, visual magnitudes give a false impression. At A0, the supergiant correction is only −0.3 magnitudes, but at B0 it is −2.7 magnitudes, and at O3, the earliest subclass now recognized, it is −4 magnitudes (see Table 12.1 ahead for a complete summary). Thus in total energy output HD 93129A is 0.3 magnitudes brighter than Cyg OB2 No. 12, and more than 1^m more luminous than the brightest M supergiant: it may be the most brilliant star in our Galaxy. Similar stars are seen in the Tarantula Nebula in the Large Magellanic Cloud (Figure 10.15 below).

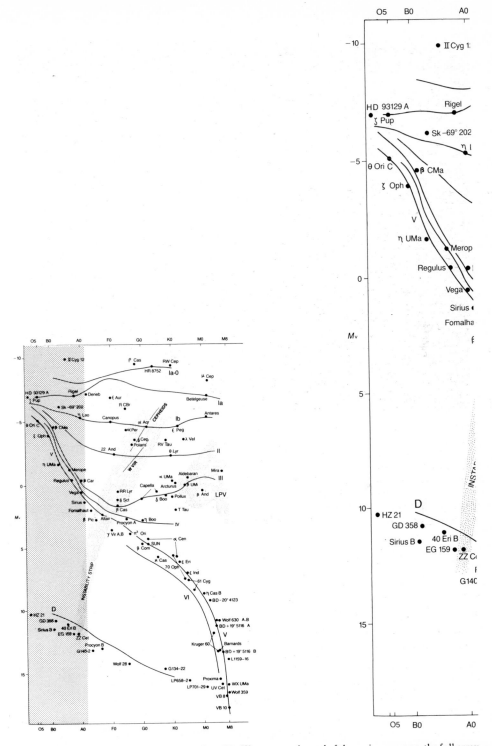

Figure 10.2. The HR diagram for classes O and B. We now see the end of the main sequence, the full array of white dwarfs, and the brightest known star, HD 93129A. This O3 Ia supergiant is actually more luminous than Cyg OB II No. 12, but appears fainter at visual wavelengths because so much of its energy emerges in the ultraviolet, i.e., its bolometric correction is very high. See Figures 3.8 and 12.7 for credits.

In the most spectular of their apparitions, the O main sequence provides us with the majority of *supernovae*. The O dwarfs are too massive to evolve into the white dwarf state, and their only fate seems to be to explode violently, giving us such marvelous objects as the Crab Nebula and the Cygnus Loop, and the weird neutron stars (Section 8.9) called pulsars, which we will examine in Sections 11.9 and 12.11.

In spite of their rarity, the O stars yield many examples of extreme behavior and are heavily observed and studied. The speed of their evolution, and their ability to process and transmute matter, give them the utmost importance in the evolution of a galaxy.

10.1 Spectra

Like type M, classification at O is open-ended. As we find hotter stars, we may adopt lower numerical subtypes. The original system began at O5, which corresponds to temperatures of about 40 000 K. We have now pushed discovery to O3 at close to 50 000 K.

Unlike other classes, we cannot discuss O star classification without consideration of emission lines. These stars are so energetic and luminous that a great number are losing enough mass to create substantial circumstellar shells that generate these features. O stars with hydrogen emission may be called Oe, but those with bright lines of He II at $\lambda 4686$ and N III at $\lambda 4634$ and $\lambda 4640$, such as ζ Pup above, are much more significant and are known as Of (see Section 3.4 for the historical development). We actually see a continuous sequence between Of and pure absorption O, and intermediate types have been introduced: O(f) for stars with N III emission and no $\lambda 4686$ at all, and O((f)) for those with the nitrogen lines still bright, but with $\lambda 4686$ now dark.

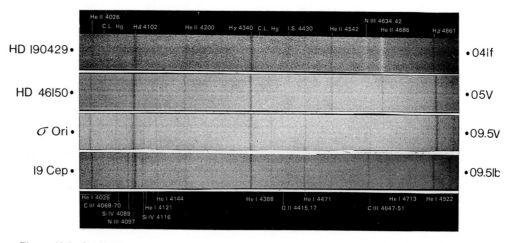

Figure 10.3. Comparisons among dwarf and supergiant spectra. Note the fading of He I, and the strengthening of He II absorption toward earlier subtypes. H, though slowly weakening, is still strong. The late O supergiant is characterized by a strengthened Si IV line. In the early O supergiant, $\lambda 4686$ of He II has turned to emission, making the star Of. From *An Atlas of Representative Stellar Spectra*, by Y. Yamashita, K. Nairai, and Y. Norimoto, University of Tokyo Press, Tokyo; John Wiley & Sons, New York, 1978.

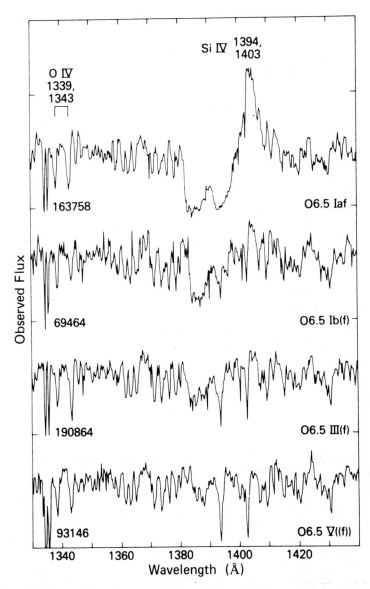

Figure 10.4. Si IV and stellar luminosity. The dwarf shows a pair of simple absorption lines. As we proceed to brighter stars, they strengthen, and in the supergiant, P Cygni emission-absorption features become very powerful. These lines show that an expanding envelope, or wind, develops toward higher luminosity. The spectra were taken with the International Ultraviolet Explorer (IUE) satellite operated by NASA. From an article in the *Astrophysical Journal*, by Nolan Walborn and Robert Panek.

In type O, of course, the He II absorptions become more prominent, as the H lines (still stronger than He II) slowly fade, and He I gradually disappears (Figure 10.3). The subtype is most commonly based upon the He II to He I ratio, although the high-excitation Si IV lines can be useful for the giants and supergiants. We must not use the prominent He II λ4686 feature, as it can be in emission, but instead employ the *Pickering Series* of that spectrum, all of whose lines arise from the fourth orbit, and are always in absorption.

Absolute magnitude can be discerned from Of characteristics that appear and develop as we move upward in the HR diagram toward brighter stars. Put another way, the λ4686 He II absorption feature is present at lower luminosity and exhibits a powerful negative luminosity effect, disappearing altogether as it is converted into an emission line (compare the top two strips in Figure 10.3). The hydrogen lines are no longer very good indicators.

Recall from Section 9.1 that the Si IV lines, which include both the optical pair at λ4089 Å and λ4116 Å and those in the ultraviolet at λ1394 Å and λ1403 Å, exhibit a notable positive luminosity effect among the B stars. This behavior, modified by wind phenomena that produce P Cygni lines, continues into middle O, where the lines are still very useful as luminosity indicators (see the accompanying set of O6 spectra in Figure 10.4.) With increasing temperature these features tend to fade, and they then become viable diagnostics for the determination of Draper type, especially among the more luminous stars where they are already strong.

10.2 Wolf–Rayet spectra

At the extreme, certain hot stars exhibit numerous broad emission bands. These are the *Wolf–Rayet*, *WR*, or *W* stars (Section 3.4, Table 3.3, and Figures 3.5 and 10.5). They are losing mass so furiously that about all we can see is the spectrum of the lower density outflowing gas.

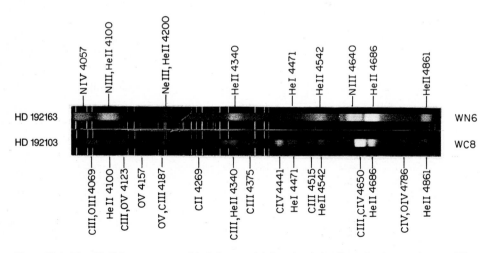

Figure 10.5. Two Wolf–Rayet spectra, with their powerful, broad emission lines of carbon or nitrogen. The upper and lower strips show the spectra of HD 192163 (WN6) and HD 192103 (WC8). University of Michigan photo, from *Astrophysics: Nuclear Transformations, Stellar Interiors, and Nebulae*, by L. H. Aller, Ronald Press Co., New York, 1954.

These stars are divided into two parallel sequences, one with prominent carbon emission (*WC*), the other with nitrogen (*WN*). Although the WC stars show some nitrogen emission, and the WN some carbon, the partition is quite clean and clear: there are no intermediate cases. Both types exhibit helium with little or no hydrogen, and the WC include oxygen in their spectra. The lines are generally in emission; very few stars show any evidence for absorption features.

We believe that we are actually looking at fundamental chemical differences, wherein the stars are either carbon or nitrogen rich. Each type is subdivided, WN3–WN8 and WC5–WC9, based empirically upon high-excitation features, for example the presence and strength of N V at early WN, N III at late WN, and the C IV to C III ratio for the WC variety. The temperatures of the WR stars are highly uncertain, since we are dealing with gas under peculiar conditions. Emission line analyses gives values of 30 000 to 50 000 K, but the continuum yields much lower numbers, around 20 000 K. The numerical subclasses are not parallel with their type O counterparts, nor are they consistent with one another. A vast amount of theoretical work remains to be done on the processes involved in the formation of these spectra.

We also see what appear to be intermediate types between normal O stars and the Wolf–Rayets. They have the usual O absorption spectra but are anomalously abundant in carbon or nitrogen, and are consequently called *ON* or *OC* stars.

10.3 Mass loss

Stars at both extremes of the Draper system exhibit powerful gas outflows. We have seen that M giants and supergiants are frequently enmeshed in thick, dusty cocoons of their own manufacture. And as noted above, the Of and the Wolf–Rayet stars show emission spectra characteristic of similar mass loss, which is driven by their own great luminosities. The major impetus to the wind appears to be pressure of the stars' powerful emission lines, which are absorbed by the outward-bound gas.

The Of varieties display strong P Cygni line profiles in the ultraviolet, as do the WR types in both the UV and the optical. A photographic spectrum of a P Cygni line, with an explanation of its formation, can be found in Section 9.5 and in Figures 9.9 and 9.10. These lines are graphically illustrated here in the spectra presented in Figure 10.4, which illustrate the luminosity dependence of Si IV in the O6 stars, and in the high dispersion IUE spectrum of HD 93129A shown in Figure 10.6. There we see powerful lines of N IV, C IV, and N V. The maximum speed of the outflowing gas (called the *terminal velocity* or v_∞) can be found by determining the wavelength of the shortward edge of the displaced absorption line and applying the Doppler formula (Section 2.12). Typical outflow velocities for the stars discussed here are 2000 to 3000 kilometers per second. From the spectrum of HD 93129A, we see that the absorption edge lies at about 1525 Å. The actual wavelength of the line is 1550 Å; the shift of 25 Å yields a remarkable speed of 4800 kilometers per second.

Mass loss rates are more difficult to determine. We must first be able to convert the line strengths into abundances, that is to calculate the number of ions actually producing the line. We know how to proceed quite well as long as we have absorptions

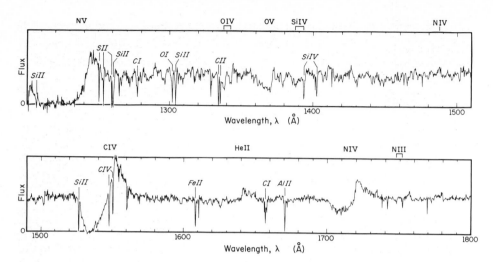

Figure 10.6. The high dispersion ultraviolet spectrum of the most luminous known galactic star, HD 93129A (O3 If). The stellar features are indicated by bold-face print. We see strong P Cygni lines of N IV, C IV, and N V. O V displays a weak P Cygni line. The Si IV lines are only a vestige of what they were in the later O stars, in response to higher temperature: all we see is a small Doppler shifted absorption. Narrow lines superimposed by interstellar gas are indicated by italic print. Adapted from the *International Ultraviolet Explorer Atlas of O-Type Spectra from 1200 to 1900 Å*, by N. R. Walborn, J. Nichols-Bohlin, and R. J. Panek, NASA Ref. Publ. 1155.

in a normal stellar atmosphere (Section 8.2) or emissions in a simple gaseous nebula (Section 10.5). But the line formation processes in expanding, high density atmospheres or coronae such as the WR stars possess are by far more complicated. They involve both collisional and complex fluorescent effects that can enhance specific lines relative to others and produce one He II line in emission ($\lambda 4686$) with others of that spectrum in absorption. Then we have to relate the ionic abundances derived to the total mass of the gas, which further compounds the difficulty, since not all relevant atoms and/or ions produce visible lines and we have to estimate their amounts. Our best estimates indicate mass loss rates as high as 10^{-5} solar masses per year for type Of, and even higher for the WR stars (compared to the Sun's minuscule solar wind value of 10^{-14}), which is very significant even in view of the short evolutionary lifetimes for these types of stars. The mass in these winds can in several instances be detected with radio and infrared telescopes as tiny encompassing clouds immediately around some WR stars. The loss rates determined from the luminosities in these spectral regions are consistent with those inferred from optical spectroscopic analyses.

With this kind of mass loss rate, we would expect to see direct visual evidence in the form of a surrounding nebula, and indeed we do. A number of the WR stars (and even a few in the Of class) have them, some quite large (Figure 10.7). They have emission line spectra, and glow by the same processes described below in Section 10.5 for H II regions. Some of these so-called *ring nebulae* are highly enriched in nitrogen and helium, confirming the similar results derived for the stellar surfaces discussed

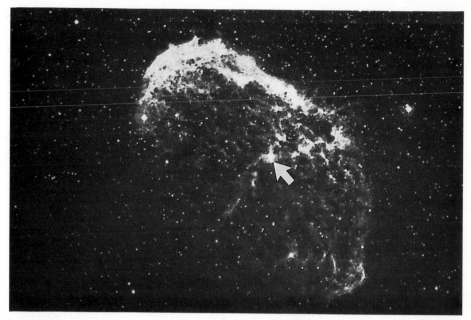

Figure 10.7. NGC 6888, a bright ring nebula in northern Cygnus created by mass flowing from the embedded WN6 star HD 192163 into the interstellar gas. The nitrogen-rich stellar spectrum is shown in Figure 10.5; the nebula is over-abundant in this element as well. The star is also surrounded by a much smaller and denser, optically unresolved cloud, detected in the radio spectrum, that consists of the outflowing matter that will eventually enter the ring nebula. A great many of these are seen in the Large Magellanic Cloud (Figure 10.14). Palomar Observatory Schmidt photograph, courtesy of H. R. Dickel.

above. They are actually formed by the outflowing gas compressing and intermixing with the surrounding interstellar medium. These curious objects may look like planetary nebulae, to be discussed in the next chapter, but the resemblance is superficial. The planetaries have an entirely different history, and were produced by lower mass M giant stars preparatory to their becoming white dwarfs.

We now believe that these high rates of mass loss can actually convert one kind of star into another, and that the Wolf–Rayet phenomenon represents an advanced stage of the Of state. There is also a suggestion that stars like the Hubble–Sandage variables that we looked at in Section 7.10, as well as others with high mass loss rates (P Cyg, η Car, S Dor) may be WR precursors. The masses of the WR stars tend to be on the low side as compared with their luminous O-type counterparts, averaging about 20 times that of the Sun. Of some significance is the large range of masses, from 10 to 50 times solar, which shows the stars' great diversity and which illustrates that they are in a variety of evolutionary states. We think that at one time they were much more massive, ordinary O stars of perhaps 50 solar masses, which have peeled away their outer layers, in the process losing large amounts of matter and exposing their inner depths, which have been enriched in carbon and nitrogen by nuclear fusion processes.

The WN and WC varieties possibly represent different evolutionary states with different internal layers exposed. We suspect that WN stars develop to WC as matter is lost in the wind, but then we should expect a difference in mass between the two and none is readily found. It is also possible that the WN and WC sequences do not follow one another and may be a result of different initial conditions such as original mass. A complicating factor is that many WR stars are components of binary systems and the close, frequently O-type, companions may powerfully influence, or even drive, mass loss. No matter what the final theoretical scenario, these stars and the process of mass loss are clearly of considerable significance to the study of both stellar and galactic evolution.

10.4 Associations

Before proceeding further, we should note again (see the introduction to Chapter 9) that a dominant fact of O star life is their tendency to be found with the early B stars in loose groupings known as *O* or *OB associations* (Figure 10.8). Typically, these systems contain a few dozen stars, the majority of type B, spread over a volume that might be as small as an ordinary cluster, or as large as a few hundred parsecs across. They generally have galactic clusters that also contain O stars at or near their centers. The naming of these assemblies is rather confusing. Some carry NGC numbers or other cluster names. For many others, an older and commonly used scheme employs a Roman numeral attached to the constellation name, such as II Persei, which contains ξ (O7e), ζ (B1 Ib), and o (B1 III) Per. This system of nomenclature was revised in the mid 1960s, so that II Persei became Perseus OB2. One must use caution since the numbers do not always stay the same, for example I Cygni became Cyg OB3. The best known of these OB groupings are probably the huge, nearby Scorpio–Centaurus association, not called by any other name, and Orion OB1, which was discussed briefly in the last chapter, and which includes the belt stars.

Associations are young, ephemeral, and related to stellar birth. Unlike galactic clusters, they are not gravitationally bound together, and are seen to be expanding and dispersing. The largest are the oldest. The young compact groups are frequently found in proximity with the birthing clouds of interstellar gas and dust: Perseus OB2, for example, is related to the well-known California Nebula, Monoceros OB2 (Figure 10.8) to the Rosette Nebula, and Mon OB1 (NGC 2264) to a huge complex of bright nebulae some 5° north of the Rosette. Wolf–Rayet stars are also included in these assemblies, which provides us with an additional evolutionary connection with the O stars other than a similarity in temperature and luminosity.

10.5 Diffuse nebulae

Now that we have established the membership of the extended O-star family, let us look in more detail at one of their loveliest and most fascinating manifestations, the diffuse nebulae, or H II regions. The best known, of course, is M42 of Orion's sword (Figure 10.1), but many other favorites come to mind immediately: the Lagoon, M8; the Trifid, M20; the Omega, M17; the Rosette (Figure 10.8); the North America,

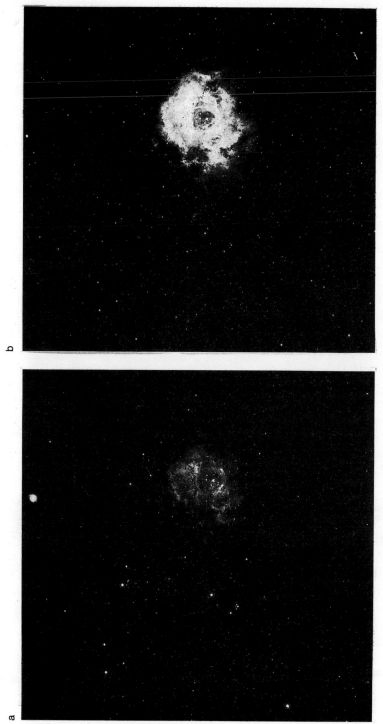

Figure 10.8. The Monoceros OB2 Association. The photos in (*a*) and (*b*) were taken in blue and red light, respectively, and the concentration of hot stars is quite obvious. The cluster, NGC 2244, at the center of the Rosette Nebula, is part of the system. The nebula is brighter in the red photo because of Hα radiation. The bright star at the top is 13 Mon, an A0 Ib supergiant. ©1960 National Geographic–Palomar Observatory Sky Survey, reproduced by permission of the California Institute of Technology.

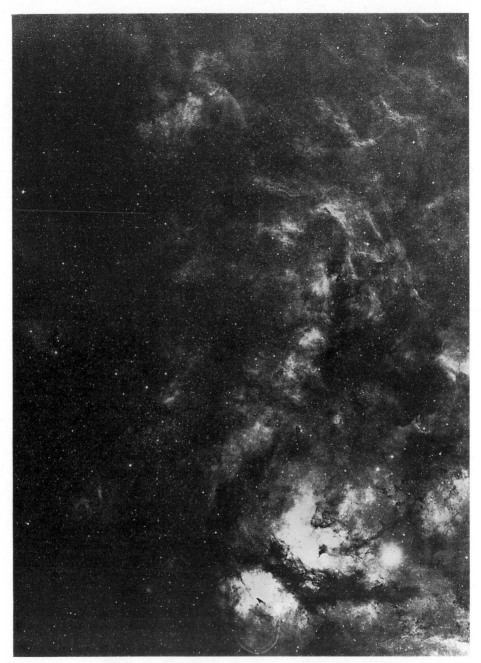

Figure 10.9. HII regions in Cygnus. The radiation from large numbers of O and B stars illuminate great sheets and clouds of interstellar matter in the galactic disk. Gamma Cygni (F8 Ib) is near the lower right-hand corner. © 1960 National Geographic–Palomar Observatory Sky Survey, reproduced by permission of the California Institute of Technology.

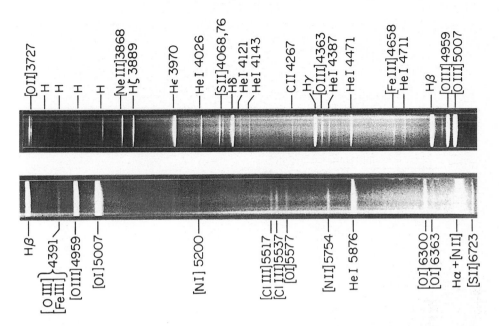

Figure 10.10. The bright-line spectrum of the Orion Nebula in the blue and red. Since these are prism spectrograms the wavelength scales are not linear, as they are for spectra taken with modern grating devices (see Section 2.13). Forbidden lines (Section 10.7) are denoted by square brackets, and the emission continuum is produced mostly by atomic processes, particularly the two-quantum mechanism (Section 10.6). Lick Observatory spectrograms, from an article by A. B. Wyse in the *Astrophysical Journal*.

NGC 7000 (Figure 11.1); the η Carinae Nebula (Figure 11.12); all lying in the star-forming plane of the Milky Way (Figures 1.7, 4.1, 9.1, and 10.9).

These nebulae all glow with emission-line spectra produced by fluorescence from the ultraviolet light of the O (and occasionally very early B) stars embedded within them (Figure 10.10). Think of a neutral hydrogen atom residing in the gaseous cloud a good fraction of a parsec from the illuminating star. The gas density is very low, perhaps only 100 to 1000 atoms per cubic centimeter (comparable to the best vacuums produced on Earth, but still very high compared to the average interstellar value of only one atom per cubic centimeter). Consequently, exciting collisions between atoms are infrequent, and the atom will have its electron in its ground, or bottom, orbit. As a consequence of its high temperature, an O star will radiate a considerable quantity of energy shortward of the critical limit at 912 Å (the Lyman limit: Section 2.7) required to ionize the atom from this lowest state. Eventually, our atom will be struck by such a photon, and will lose its electron. Follow the passage of the electron in Figure 10.11.

After a considerable time in space, in which it exchanges some of its energy with other electrons, perhaps losing some by close passage to protons (other ionized H-atoms), and losing still more through collision with heavy ions (see below), the freed electron will finally encounter another proton head on, and the two will join, or recombine through an electrostatic interaction. But whereas the electron arose from the

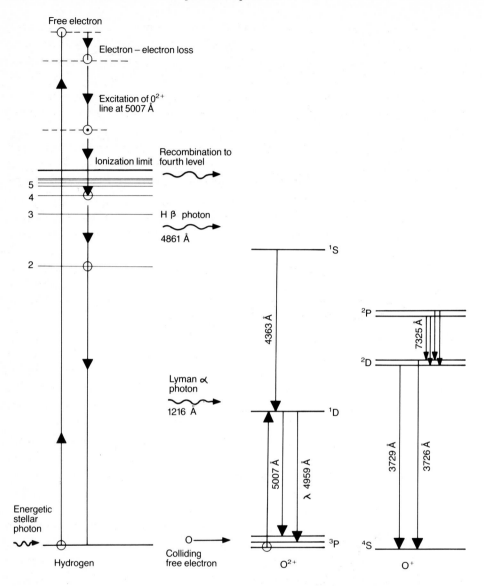

Figure 10.11. The formation of the nebular spectrum. The illumination of a diffuse nebula starts with the photoionization of hydrogen by starlight. After losing some energy to other electrons or to nearby passing protons, the freed electron collides with an O^{2+} ion, sending the bound electron upward from the energy state '^{3}P' into the one labeled '^{1}D', and losing some of its own energy in the process. (The names of the levels are derived from the rules of quantum mechanics: see Section 2.9.) It then recombines with another proton. Here we show it landing on the fourth level as it produces energy in the Brackett continuum (Section 2.7), then jumping to the second with the creation of an Hβ photon, and finally to the first, producing the Lyman α line at 1216 Å in the ultraviolet. The electron in the O^{2+} energy state called '^{1}D' is shown dropping to either of the two top sub-levels of '^{3}P', whence it radiates light at 5007 Å or at λ4959 Å. These lines are very prominent in Figure 10.10. Other possible transitions are shown for both O^{2+} and O^+. The scales of the oxygen-atom diagrams relative to that of H are doubled for clarity, and sublevel separation is enormously exaggerated. Diagram by the author.

ground state, it may now land in any orbit at all. It might jump directly onto the bottom level, releasing a far-ultraviolet quantum of energy, or it may pass through intermediate orbits on the way down, radiating at a variety of wavelengths. If it happens to pass from 3 to 2, it produces Hβ; from 4 to 2 (as in Figure 10.11), Hβ; and so on. Thus when we look at the Orion Nebula's spectrum (Figure 10.10), the H lines appear strictly in emission. Hα is by far the most likely to be produced, and is the strongest, which is why diffuse nebulae usually appear red in color photographs.

The competing rates between this *photoionization* and the *radiative recombination* are such that the gas at any given time is in a high state of ionization, and consists largely of a sea of free protons and electrons called a *plasma*; fewer than one percent of the atoms are neutral. The star (or stars) will fully ionize the surrounding gas out to a radius at which all of the ultraviolet photons are absorbed. The cloud will at that point, which may be several parsecs from the center depending on the UV flux and gas density, suddenly become neutral. We will then see the nebula as a spherical bubble set within the surrounding neutral medium. The theory was first outlined by Bengt Strömgren, hence the term *Strömgren sphere*, used earlier in Section 9.7. The Rosette Nebula, seen in Figure 10.8, provides an excellent example. Of course the gas distribution is usually anything but uniform and homogeneous, so that the nebulae become quite distorted and are often far from symmetrical.

The concept of temperature loses some of its meaning for these gases, much as it does in the solar corona where the gas density is also very low. We relate the temperature of a nebular gas only to the velocities of its atoms and electrons, so that we define what we call an *electron temperature*. Typical values are about 10 000 K, that is, the atoms are moving at speeds typical of those found in the atmospheres of A stars. But because the nebulae are not blackbodies, they do not radiate at a level appropriate to this temperature, and appear very much fainter, analogous to the contrast between the dim million degree solar corona and the brilliant 6000 K photosphere (Section 6.4).

10.6 The nebular continuum

Note also in Figure 10.10 the presence of a faint continuum, which can be better seen in the spectra of planetary nebulae (Figure 11.4), which we will examine in the next chapter. Some of it in the Orion Nebula's spectrum may be due to faint background stars, but a component is produced by the atomic processes of the nebula itself. Part of the nebular continuum is caused simply by the free-bound (recombination) and free–free processes that we introduced in Section 2.7. The free–free continuum is easily observed in the radio spectrum.

Most of it, however, is produced by the *two-quantum* mechanism. Usually, an electron in hydrogen's level 2 will simply jump downward to level 1 with the production of the Lyman α line at λ1216 Å. However it is also possible that *two* photons can be emitted simultaneously. The sum of their energies must equal that of Lyman α, but otherwise there are no restrictions; they can be of equal energy, or one may be high, the other low. The result is a continuum that pervades the spectrum with a limit at λ1216 Å. Since a Lyman α photon, once created, will suffer numerous successive re-absorptions and re-emissions as it tries to work its way out of the nebula, the odds of it breaking down into two photons become reasonably good, and a strong continuum is produced. A similar process in ionized helium will produce additional continuous radiation in the spectra of planetary nebulae (the exciting stars of H II regions are too cool to produce He^{2+}) down to He II Lyman α at $\lambda 304$ Å.

10.7 Forbidden lines

Many other kinds of atoms produce similar types of recombination emission, most notably helium. But some of the strongest lines cannot be explained in this manner, and even defied identification with known ions until about 1930. These are the *forbidden lines*, so called in opposition to the *permitted* spectra described in Section 10.5; under ordinary laboratory conditions they have such a low probability of formation relative to other processes that they cannot be seen. But in nebulae, with their low densities and high masses, circumstances are just right.

The best known forbidden lines are those of doubly ionized oxygen, O^{2+}. The ordinary permitted lines are called O III, and the forbidden variety are enclosed in brackets, e.g. [O III]. Located just above the bottom orbit or energy level (which itself is a triple set) are a pair of what we call *metastable states*: again see Figure 10.11. These possess quasi-ground conditions in which an electron will linger for a long time. Electrons that are bound into the lowest states of O^{2+} are knocked upward into these metastable levels through collisions with the free electrons of the gas. Even though their lifetimes in the metastable states are long (say 10 or even 100 seconds, as opposed to 10^{-8} second for a hydrogen upper level), they will eventually drop downward, producing powerful lines at $\lambda 5007$ Å and $\lambda 4959$ Å and a weaker one at $\lambda 4363$ Å (see Figure 10.10). The two bright longer-wave [O III] lines dominate most nebular spectra and can give some nebulae a greenish cast. Well before the discovery of their origin, they were thought to be produced by an element called 'nebulium' that had not yet been found on Earth; they are sometimes still called 'N1' and 'N2', not to be confused with the symbol for nitrogen. Other forbidden emissions, such as [Ne III] and [O II] (from O^+, also illustrated in Figure 10.11), are easily detected in the ultraviolet, and at longer wavelengths we find pairs of [N II] and [S II] lines near $H\alpha$. The intricate mixture of permitted and forbidden lines is nicely displayed in the Orion Nebula's spectrum shown in Figure 10.10.

Nebular line spectra are simple to analyze, and thus we can use them to great advantage to determine the temperatures, densities, and chemical compositions of the observed clouds. And the atoms we detect are often just those (for example helium, oxygen, and neon) for which it is most difficult to derive atomic abundances from stellar absorption lines. Study of the nebulae, and the examination of the variation in chemical composition within the disk of our Galaxy and across the faces of others (Figure 10.12), allow us to probe such significant subjects as galactic evolution and the mechanism of star formation.

Figure 10.12. The famous Whirlpool galaxy, M51. The knotty structure of the spiral arms is caused by individual clumps of O and B stars and their associated H II regions. Spectroscopic examination of such galaxies provides us with data on the variation of chemical composition with position, and hence on the evolution of the system. There is evidence that the metal composition drops outwardly in the disk as a consequence of long-term galactic evolution. The maximum stellar temperatures then increase outwardly in response to lower atmospheric opacities, which shrinks the stars and make them a bit hotter. National Optical Astronomy Observatories (Kitt Peak) photo.

10.8 Interstellar lines

The H II regions are but one way in which the O stars show us the interstellar medium. Starlight must pass through the interstellar gas on its way to us, and in principle, the dark clouds will superimpose their absorption spectra onto those of all the stars. Many stars, however, are not behind enough dark matter to allow the lines to build up much strength. More importantly, these absorption features come from states of low ionization and excitation, and are usually lost within the similar and overpowering complex spectra of later-type stars. But for O and B stars the competing low-energy features have disappeared, so that they have very simple spectral backgrounds against which the interstellar absorption features show quite well; moreover the stars are luminous and distant, and are in the galactic plane mixed in with a great deal of interstellar gas.

The first detection of such a feature was made long ago in 1904, in the spectrum of δ Orionis. This B0–O9 pair are in rapid orbit about one another and the stellar lines are Doppler shifted back and forth over a 5.7 day period. But the Ca II K line stays at a constant position, and must therefore be formed by an intervening medium. Additionally, Ca II is not intrinsic to early B stars, and the line is anomalously narrow compared to the broadened stellar absorptions (Figure 10.13).

Other optical lines such as Fraunhofer D of Na I, and molecular absorptions caused by CN and CH$^+$, are identified as well. At very high dispersion, the features are frequently Doppler-split into multiple subcomponents that allow us to study the structures and dynamics of individual clouds.

We have also obtained extraordinarily good data on interstellar lines from examination of the ultraviolet spectrum between λ1000 and λ3000 Å by such Earth-orbiting telescopes as *Copernicus* and the *International Ultraviolet Explorer*. Here we detect numerous dark interstellar features that arise from many elements, such as the CNO group, silicon, and common metals (Mg, Al, K, Fe, Ti, Ni, Cu, Zn, Mn) as well as from molecular hydrogen; see the spectrum of HD 93129A in Figure 10.6 where several are marked. The studies show a remarkable depletion of certain kinds of atoms,

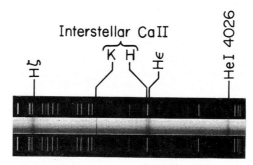

Figure 10.13. The near-ultraviolet spectrum of λ Orionis (an O8e + B0.5 V binary), the star that helps form Orion's head. The narrow H and K lines are not formed in the star, but by the intervening interstellar medium. Kitt Peak National Optical Astronomy Observatories (Kitt Peak) Spectrogram, courtesy of E. C. Olson.

notably those heavier than sodium. Apparently, these elements, in addition to those such as carbon, have been drawn out of the gas in the formation and/or growth of interstellar grains.

In addition to their dimming and reddening effects, the solid grains also are responsible for superimposing *diffuse interstellar features* onto the stellar spectra. The best of about twenty of these falls at λ4430 Å. Because the atoms or molecules that produce them are in some way locked into the dense solids the lines are several Ångstroms wide, and are usually quite shallow. However their origins elude us: they might be produced by atoms in the crystal structure of the grains themselves, or by molecules attached to their surfaces.

Most of what we know about the gases of the interstellar medium comes from the radio spectrum, where we see signatures of numerous chemical compounds, which range from the simple carbon monoxide through formaldehyde and alcohols to complex and exotic carbon chains. These data coupled with what we observe in the infrared and optical, allow us to take our first real steps toward a genuine comprehension of the turbulent, almost chaotic, interstellar medium. The knowledge of how stars are formed is then within our grasp.

10.9 The end of the main sequence

We have, over this course of chapters, constantly alluded to the drop in the absolute numbers of stars as we proceed toward earlier types. These figures are collectively known as the *luminosity function* (which is often generalized to include giants and supergiants). Seventy-two percent of all dwarfs are of type M; then 14, 9 and 4% are of type K, G, and F respectively. Class A constitutes one percent, then the numbers simply plummet: only 0.1% are B stars, and the O types contribute an amazingly minuscule 0.00004% with the numbers dropping steeply from O9 to O3. The various subsets of this extreme class, especially the WR varieties are even rarer. This run of percentages is totally contrary to our naked eye experience. The M dwarfs are so faint that we see none at all with the unaided eye, and the luminosities climb so rapidly that they offset the diminishing numbers, allowing the sky to seem to be dominated by B and A stars. Though the members of class O are more luminous yet, their extreme rarity still makes them visually scarce.

Quite obviously, nature prefers to create low mass stars. We do not yet even begin to know the reason. The lowest mass M dwarf has about 8% the bulk of the Sun: below that, thermonuclear fusion cannot be sustained. Now, what about the other end? What are the most massive stars that can be created?

We are still not sure of the answer, but they are almost certainly to be found among the brilliant, rare O3 stars. Mass measurements are very difficult to make, but our best estimates, derived from theory and based upon observed luminosity and temperature, hover around 120 solar masses. Included in this list are HD 93129A and HD 93250, both lodged within the extraordinary η Carinae nebula (see Section 11.7). Also included are the stars that light the gigantic Tarantula Nebula (30 Doradus) of the Large Magellanic Cloud (Figures 10.14 and 10.15). Near the center is an unusual 'star'

Figure 10.14. The Large Magellanic Cloud, which contains a great number of hot stars and diffuse nebulae. The Tarantula Nebula, 30 Doradus, lies at left center. Harvard College Observatory photograph.

Figure 10.15. A close-up view of 30 Doradus. The nebula is roughly 200 pc in diameter, 30 times the size of the Orion Nebula, which at that distance would appear only the size of one of the larger stellar images. The gas is ionized by a remarkable cluster of O and WR stars at its core. National Optical Astronomy Observatories (CTIO 4-meter) photograph, courtesy of H. Spinrad.

called R (for Radcliffe) 136a, or HD 38268. For a brief time it was thought to be a single superstar of as much as 1000 solar masses, but we have recently learned that it is a remarkable, tightly compacted cluster of early O and WR stars. We see a few other of these giant H II regions like 30 Dor, presumably lit by similar stars, sprinkled about in other galaxies, for example the great NGC 604 in the Triangulum spiral M33.

It is doubtful that we will find any stars much more massive than those of class O3. The formation process at the upper end of the main sequence is self-limiting. The winds that develop as we proceed upward on the HR diagram become even fiercer as the great luminosities that drive them increase. At some point above 100 solar masses so much radiation would be produced that the star would immediately cut itself to a smaller size. The greatest stellar mass to be found in a galaxy is probably a matter of chance. The larger the galaxy and the more stars it contains, the greater will be the odds that an exceedingly rare ultramassive star could be formed. But even if such an event were to occur we would have to find such a remarkable body before its outer layers dissolved away under its extraordinary wind. So far, the neighborhood of 120 times the mass of our Sun seems to be the common limit.

With these stars we think we have at last seen the top of the main sequence. But the members of the early O classes are so rare and distant, and the theory required to interpret them is so difficult, that they epitomize our still vast ignorance of the workings of stars and how they are made. Our subject, begun a hundred years ago with the first attempts at spectroscopic analysis and classification, is still remarkably wide open for exploration.

11 *Extraordinary classes*

In the summer of 1975 a bright new naked-eye star flared into view in northern Cygnus. For a few days it appeared that the Swan had two tails, so brilliant was the visitor (Figure 11.1). However, the disruption was to be short, and the constellation returned to its normal appearance within about a month. What we had witnessed was a *nova*, now known to be a surface eruption on a white dwarf. Every few decades one will occur that is bright enough to reach first or second magnitude and change the sky awhile, such as the Novae Persei 1901, Aquilae 1918, Herculis 1934, and Cygni 1975. They are much

Figure 11.1. Nova Cygni 1975. This 'new star', also known as V1500 Cygni is indicated by the arrow, and rivaled Deneb, labeled 'α'. Searches of Palomar Sky Survey prints showed no star at that position brighter than the 21st magnitude, which makes this nova the most energetic ever observed, with an extraordinary brightness increase of over 19 magnitudes. The North America Nebula, NGC 7000, is at the center. Photograph taken Aug. 31, 1975, courtesy of A. E. Morton.

Figure 11.2. A favorite planetary nebula, the Dumbbell, M27. The central star of this well-evolved object is at the exact center of the nebular image, and is one of the hottest known, with a Zanstra temperature (Section 11.2) of 125 000 K, and a luminosity 100 times the Sun's. Photograph courtesy of David Healy.

more common than this, however; one will reach naked-eye status every few years, and the telescopic variety is at least an annual event. They are common enough that by 1928 they warranted their own spectral class, Q, which was borrowed from its original usage wherein it designated 'all other spectra' (Table 3.2).

The other of the two remaining Draper types that we shall consider here, P, is unique in that there are no naked-eye examples. These are the *planetary nebulae*, which exhibit extensive emission-line spectra. In their more highly evolved forms they are among the loveliest of non-stellar objects, with the famous Ring Nebula in Lyra (M57) and Vulpecula's Dumbbell (M27, Figure 11.2) as archetypal examples. The latter is very prominent, and is easily visible with binoculars.

The P and Q designations were long ago dropped from active use, but they provide a springboard for us to examine modern classification and the current state of knowledge about these highly evolved objects. We shall also use this opportunity to look at one of the sky's oddest stars and at other strange ones that exhibit combinations of different spectral systems.

11.1 The planetary nebulae

These remarkable objects received that appellation in the 18th century from William Herschel, as they reminded him of the small disks of the planets as seen through a modest telescope. They of course have nothing whatever to do with planets or planetary formation, and are actually shells of gas produced by dying stars. Their gaseous constituency was discovered by William Huggins in 1864, when he observed the spectrum of NGC 6543 in Draco and found emission lines.

The spectra of planetary nebulae are usually a composite, a mixture from the emitting gas and the illuminating central star. The stellar component may contain either absorption or emission lines or both. Most planetaries are sufficiently large or bright enough to enable us to isolate the pure nebular spectrum. The same cannot be

Figure 11.3. Not all planetaries are large. This one, NGC 40 in Cepheus, is 30 seconds of arc across, and is also known as HD 826. The central star, of type WC 8 (Section 11.3), is one of the coolest known, with a Zanstra temperature of about 30 000 K. It was originally classified as Pf on the basis of its stellar He II line. University of Illinois Prairie Observatory photograph.

said for that of the embedded star, however, which is frequently contaminated by the radiation from the surrounding gas, and consequently is often quite difficult to observe with precision.

Let us look first at the nebulae themselves, and then at the stars that power them. These objects are gaseous spheres and shells ejected by distended red giants, and are the remnants of the mass-loss processes that convert giant stars into white dwarfs. Since the nebulae are produced by outflowing winds, they are in a state of steady expansion. As we would then expect, we see a great range in radius, from stellar, or unresolved, to huge objects angularly comparable to the Moon. At the measured average expansion rate of about 20 km/s, the observed lifetime of a nebula is roughly 35 000 years before it reaches a radius of 0.8 parsec, at which point it becomes faint enough to avoid easy detection. The Ring and Dumbbell are both intermediate-sized objects with radii of roughly 0.2 parsec. Amateur observers are generally unaware of the large number of very bright, small objects (e.g. Figure 11.3) that are only a few seconds of arc (and a few hundredths of a parsec) across. These are the bright ones found in the HD catalog that define class P; the stars by themselves are generally too faint for the original classifiers to have seen.

The mechanism for the origin of the nebular spectrum is exactly that described earlier for the diffuse nebulae in Sections 10.5 through 10.7. But the planetaries have a much greater range in excitation and ionization (Figure 11.4), reflecting a similar comparative spread in the stellar temperatures. Galactic H II regions never exhibit

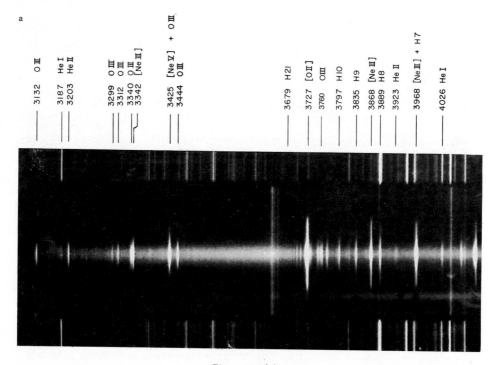

Figure 11.4 (a).

b

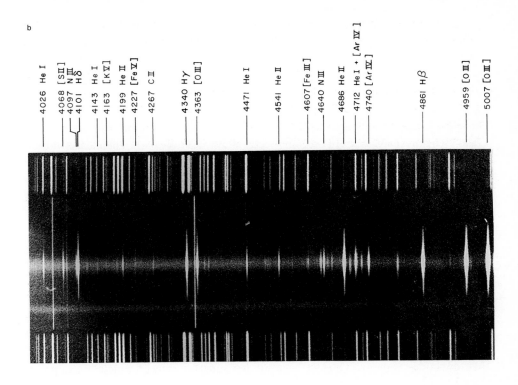

c

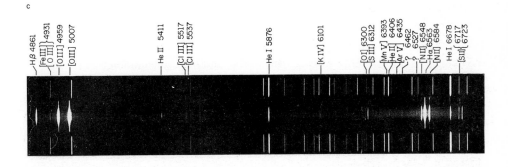

Figure 11.4. The spectrum of the remarkable planetary NGC 2440 in Puppis, from the violet atmospheric limit to the red. Many high excitation lines are visible. The very faint nucleus, at 220 000 K, is the hottest known. The emission continuum is produced by atomic processes in the nebula (Section 10.6). The spectrograph slit was set along the major axis of the object, and considerable structure can be seen, especially in the red. The long, very uniform lines are from mercury-vapor street lamps near the observatory. Lick Observatory spectrograms: (a) and (b) from an article in the *Astrophysical Journal* by L. H. Aller, S. J. Czyzak, and J. B. Kaler; (c), courtesy of L. H. Aller.

HeII lines (which require the capture of an electron by doubly ionized helium for their formation), but these emissions are the rule for planetary nebulae. At one time class P was subdivided Pa through Pf, roughly in order of increasing excitation, but the system long ago fell out of use.

Also unlike the diffuse objects, which are found strictly in the galactic disk, planetaries are a mixture of population types (Section 1.8), and although they concentrate toward the plane of the Milky Way, they are also found very far into our Galaxy's halo. Analyses of H II region spectra allow the study of the chemical evolution only of the galactic plane, whereas a similar examination of planetaries yields analogous information not only for the disk itself, but perpendicular to it into the ancient galactic halo. These objects therefore allow us to probe far into the past, to earlier days of galactic development. What we see from both kinds of nebulae is the same sort of increase in the abundances of heavy elements with time as we find for the stars.

An additional fundamental difference between the two types of emission nebulae is that planetaries sometimes show the influence of the stars that created them. Some exhibit very significant enhancements in helium, nitrogen, and carbon relative to the abundances found in the interstellar medium. We are seeing the by-products of nuclear burning that were cycled from the hot core into the outer envelope of the star before it was to be lifted away to form the nebula, the same process that developed the abundance enrichments in the carbon stars that we examined in Section 4.7: these planetaries are their direct descendants. Thus we have another superb means by which we can probe the interior of the original star, and examine theories of stellar structure and evolution. The enriched material enters the interstellar medium, and is a source of heavier elements that contributes to the chemical evolution of the Galaxy.

11.2 Planetary nuclei

The central stars of the planetaries are what remain of the giants after the mass-loss processes that produce the nebulae. They are little more than the old nuclear burning cores of these once-mighty behemoths, and have in common that they are all very hot (otherwise no nebulae would be visible), and that they are all becoming white dwarfs. The O stars, with temperatures of up to 50 000 K, have been presented as the hottest class, but that extreme pertains only to the main sequence. The planetary nuclei are in the throes of heir final contractions, and therefore can become much hotter, with the highest observed values commonly about 125 000 K. It is really not proper to plot them on a standard HR diagram, since because of their extreme, often odd, spectra and very high temperatures they cannot be assigned normal O subtypes.

The actual high temperature limit is still very uncertain. As a star becomes hotter for a given total luminosity, a greater fraction of radiation is produced shortward of the ionizing Lyman limit at 912 Å, the nebula brightens at the expense of the visual stellar magnitude (i.e. the bolometric corrections are huge), and the nucleus becomes very difficult if not impossible to detect against an overwhelmingly bright background. The central stars of some planetaries have not yet been seen, though they are surely there. The current temperature record hovers at about 250 000 K for a number of bright nebulae. These stars are barely detectable with the most sophisticated available imaging techniques. More like them will certainly be found as our observational methods improve.

In spite of the observational difficulty imposed by the nebula, it is indispensible to the study of the star and provides the principal means for temperature determination. Above about 40 000 K, blackbodies are practically indistinguishable from one another in the optical part of the spectrum. Even that part of the ultraviolet observed from space provides only limited information. And we do not yet understand the spectra well enough to use them to infer temperatures as we would for ordinary stars.

To derive temperature, we make use of the surrounding gaseous cloud as an ultraviolet photon counter. Assume that a planetary is 'optically thick' to starlight, that is, it absorbs all the stellar UV radiation, and is consequently a Strömgren sphere (Section 10.5) set into a larger volume of gas. Each photon shortward of the Lyman limit can produce one ionization in the nebula, and by a short chain of reasoning put forth by Hermann Zanstra in 1927, we see that each one of these ultimately causes one Balmer photon to be emitted. In order for the nebula to be in equilibrium (which it is since we do not see it vary), every ionization must produce a recombination. The recombining electron must land on some particular energy level. If it falls directly to the first, generating a Lyman continuum quantum, it simply recreates the ionizing radiation of the star, and we have lost nothing. If it hits the second, a Balmer continuum photon will be produced together with Lyman α and that is that. If it lands in an orbit above the second, then we have additional choices. It could go to level 2, and we see the Balmer line. It could, however go first to the ground state (perhaps after a cascade to a lower level still above number 2), producing a Lyman photon. The nebula is already defined as thick in the Lyman continuum, and it must therefore be thick in the Lyman lines as well. Therefore this photon will immediately be reabsorbed to place the electron back to its original upper level. After enough such scatterings, the electron by chance will drop to level 2 instead of 1, and once again a Balmer photon is created.

By measuring the luminosities of all the Balmer lines and continuum, or by observing Hβ alone and applying appropriate theory that relates the strengths of all Balmer radiation to that of Hβ, we can find the total ultraviolet photon output radiated by the star. We then compare this figure with the visual photon emission rate derived from a magnitude measurement, and find the value for a ratio that is exquisitely sensitive to temperature (Figure 11.5). For the hottest stars, we may apply the same reasoning to ionized helium to determine the stellar luminosity shortward of the He$^+$ Lyman limit at λ228 Å. The results are termed *Zanstra temperatures*. If the spectrum of the nucleus is truly that of a blackbody, and if all its radiation is absorbed, the two values will be equal. If they are not, we can make statements about the opacity of the gas or departures from the blackbody. With these measurements plus the distances we can also deduce the stars' total luminosities relative to that of the Sun.

What we finally find after all the analysis is that the central stars, far from being a homogeneous set, are quite a diverse lot. Temperatures range from a low of 25 000 K or so, similar to those found in early B, up through the numbers described above.

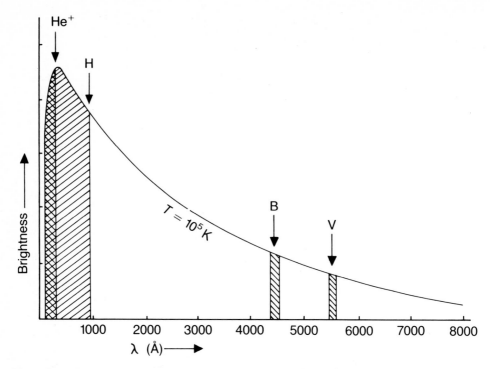

Figure 11.5. The derivation of Zanstra temperatures. The curve shows the continuous spectrum of a hot, 100 000 K blackbody that represents a typical planetary nebula nucleus. From the count of nebular hydrogen Balmer line photons we can find the total amount of radiation shortward of the Lyman limit at 912 Å. The observed B or V magnitude gives us the energy produced at $\lambda 4500$ Å or $\lambda 5500$ Å. The ratio of UV to optical radiation is a very sensitive indicator of temperature, as shown by the comparative curves in Figure 1.11. The same reasoning applied to the nebular He II lines gives us the ratio of the energy shortward of the He$^+$ Lyman limit at $\lambda 228$ Å to that at B or V, and another value of temperature.

Luminosities go from 10^4 that of the Sun, comparable to those found for supergiants, down to 10 times solar and less. From these figures, we deduce radii that range between 2 or 3 times our Sun's to those typical of hot true white dwarfs, and (from the luminosities and applications of theory) infer masses of between roughly one-half and one solar mass. It is still unclear, however, that the stars are in fact blackbodies, or are even close, so that our temperature analyses remain uncertain. A great deal of work remains to be done on understanding the construction of these extreme stellar atmospheres.

11.3 Classification of the central stars

The variegated nature of these stars is supported by the divergence found among their spectra. The bodies have a strange internal constitution: a carbon-rich core overlain by a helium envelope sometimes topped by a hydrogen skin, all the result of eons of nuclear burning. It is no surprise that the planetary nuclei possess a rich variety of sometimes bizarre spectra that are loosely correlated with temperature and luminosity. What what we actually see are stars in the process of developing their white dwarf

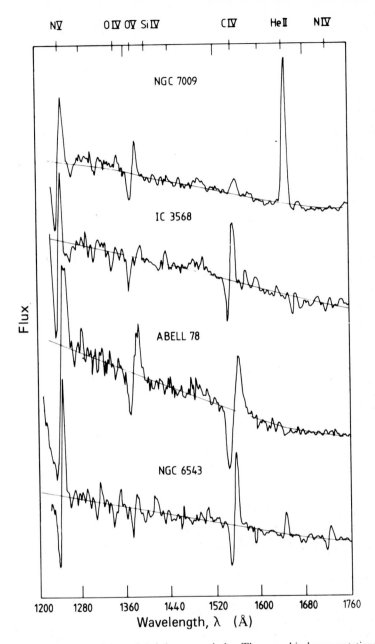

Figure 11.6. Ultraviolet spectra of the nuclei of planetary nebulae. These graphical representations, all made with the *International Ultraviolet Explorer* satellite, exhibit considerable diversity among the central stars. P Cygni line profiles of C IV, O V, and N V are especially prominent, showing that all four stars have vigorous winds. The stars, in order of increasing temperature, are NGC 6543 (45 000 K), IC 3568 (61 000 K), NGC 7009, and Abell 78 (both near 80 000 K). Note the growth of the high-excitation O V feature with temperature. The C IV P Cygni line, however, is reduced to a simple emission feature in NGC 7009, suggesting that A78, for which it is very strong, is carbon-rich. The He II line for NGC 7009 is of nebular origin, but for NGC 6543 it is produced by the star. Note also how the continuum climbs in brightness toward shorter wavelengths, as expected for blackbodies at high temperature. Panel from *Planetary Nebulae* by S. R. Pottasch, D. Reidel Publishing Co., 1984.

characteristics. Classification has long been a problem, compounded by the difficulty of observing weak stellar features against an often bright nebular background. Because of the high temperatures the spectra frequently bear similarities to those of the O stars, and for convenience we have borrowed the nomenclature. But keep in mind that the central stars simply mimic the behavior of their much more massive early-type counterparts.

Some nuclei, particularly the less luminous ones, exhibit only absorption lines. These are often termed *subdwarf O*, a name sometimes used for central stars in general even though they have little in common with traditional subdwarfs. Many of the more luminous ones are still losing mass through vigorous high speed winds and therefore display emission features much like those found in the spectra of the luminous O supergiants. Some spectra have prominent P Cygni lines (Section 9.5), especially in the ultraviolet (Figure 11.6). They are termed *Of* (Section 10.1) when He II and N III are in emission and *Wolf–Rayet, WR* (Section 10.2), when we find the appropriate broad features. These stars must *not* be confused with their classical analogues. A major difference is that all of the WR nuclei are of the carbon subtype WC, a result of the very different internal constituents of these highly evolved stars.

The planetary version of the WC stars, with numerical subgroups WC 8 through WC 10, are the coolest, below 36 000 K, with the other types spreading upward in temperature. Within the collection of the hottest stars, above 80 000 K, we find an extreme that cannot exist on the classical HR diagram: stars that display powerful lines of five times ionized oxygen near $\lambda3820$ Å (Figure 11.7). At their limit the energy radiated by O VI alone is equal to the luminosity of our Sun! These *O VI stars*, are sometimes misleadingly classified WC 2 through WC 4 as if they are only extensions of the Wolf–Rayet sequence, a link that seems quite tenuous.

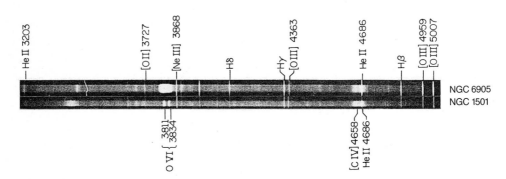

Figure 11.7. Central star spectra of the planetaries NGC 6905 in Delphinus (upper) and NGC 1501 in Camelopardalis (lower). Both are O VI stars, the highest, most highly excited type of emission-line nucleus. At temperatures of 80 000–120 000 K, they are far hotter than the 30 000–40 000 K WC stars. The broad stellar lines are identified across the bottom and the nebular features across the top. The stellar features not identified are a mixture of O III and O IV. The O VI lines of the NGC 6905 nucleus are Doppler broadened so much that they are blended together, indicating a wind with a mass outflow velocity of about 4000 km/s. The spectrum of the nebula is seen superimposed on that of each star, illustrating the difficulty that attends the study of these objects. Note the strong nebular He II lines, which suggest very high stellar temperatures. Lick Observatory spectrograms, from an article by L. E. Smith and L. H. Aller in the *Astrophysical Journal*.

We do not understand the origin of the variety of types very well, nor why winds are switched on in some stars, whereas others with similar temperatures and luminosities barely show them at all. But with modern techniques and advances in theory, we are showing that these remarkable stars, the nuclei of objects that in times past had been considered 'unimportant pathological specimens', play a fundamental role in the stellar aging process.

11.4 Novae

A nova is without question among the heavens' most dramatic phenomena: a star suddenly appears where none seems to have existed before. However its name, from *nova stellarum*, which means 'new star', rings false. In the case of almost every one of these, if we are lucky to have photographed the sky previously with a sufficiently deep exposure, or wait and obtain a picture after the outburst is done, we will find a faint star responsible for the eruption. The event does not involve the new, but the old: aged stars in special states of evolution.

Novae are extreme examples of a heterogeneous set of stars collectively referred to as *cataclysmic variables*, or *CVs*, which include the *recurrent* and *dwarf novae* as well as the *classical* version (the first kind known) that is our focus at the moment. Classical novae are characterized by a quick rise of more than 10 magnitudes over a period of roughly a day, followed by a much slower decline (Figure 11.8). Absolute visual magnitudes as high as −10 can be reached. Within that framework, they are highly individualistic. A 'fast' nova may drop from peak brightness through its first 3 magnitudes in a month, whereas a 'slow' one may take half a year. Anywhere from a year to several decades may elapse before the star recovers to normal. Generally, the brighter they are at maximum, the faster they will pass through their changes.

The light curves can also be very different from one another: some are quite smooth following peak brilliance; others develop strong oscillations after an initial drop, apparently caused by fluctuations in the rate at which mass is expelled during the eruption. Still others display broad, deep dips, followed by a recovery, produced by development of a dust shroud.

The spectrum of the developing nova (Figure 11.9) is as complex as one might expect following such a violent outburst, and is in a constant state of change that is correlated with the light curve. At maximum, we typically see a spectrum that is reminiscent of that of an A supergiant like Deneb, as the thick cloud of gas lifts off the star. Within the next few days we observe emission lines develop from the expanding, more rarefied envelope, and several sets of blue-shifted absorption lines formed by different systems of outwardly moving gas seen against the stellar continuum. These features are somewhat reminiscent of P Cygni-type lines, but are much more complicated. Maximum outflow velocities can reach 2000 km/s. Gradually, the nova and its continuum fade, and the spectrum becomes dominated by broad emission lines.

Classical and recurrent novae, as well as the dwarf versions (Section 11.6), work on a similar theme: the transfer of matter from one component of a binary system to the other. When a pair is born, more commonly than not each member will have a different

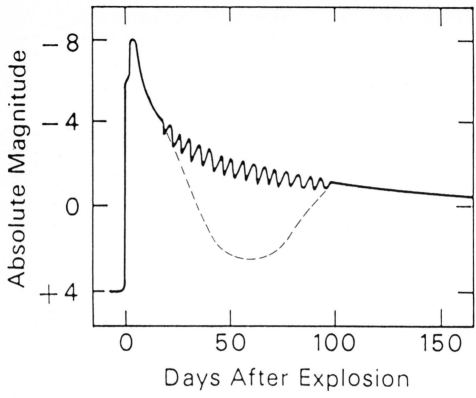

Figure 11.8. A schematic light curve of a nova. After a short pause, the exploding star reaches maximum a few days after outburst. The slow decline back to normal, which may take years, may be interrupted by oscillations or a deep depression. There are wide variations upon this theme. Adapted from an article by D. B. McLaughlin in *Stellar Atmospheres*, J. L. Greenstein, ed., University of Chicago Press, Chicago, 1960.

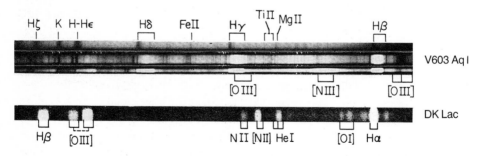

Figure 11.9. Nova spectra. The upper three strips show blue spectra taken of Nova Aquilae 1918 (also assigned the variable star name V603 Aql) near maximum, then three days and three weeks later. In the first, we see only the absorption line spectrum of the expanding ejecta. In the second, emission lines and a complex set of blue-shifted absorptions appear. The third shows further development, and is termed an 'Orion spectrum' because of its emissions. The lowest strip illustrates a typical nebular phase, with a yellow–red spectrum of Nova Lacertae 1950 (DK Lac) taken almost 6 months past maximum. University of Michigan Observatory spectrograms, taken from an article by D. B. McLaughlin in *Stellar Atmospheres*, J. L. Greenstein, ed., University of Chicago Press, Chicago, 1960.

mass, and each will evolve at a different rate (Sections 1.17, 4.8, and 5.7). The weightier of the two will be the first to develop into a giant. If the stars are very well separated, its growth will have little effect on its companion. The giant will continue to evolve, and the result will be a white dwarf orbiting a main sequence star, which in turn will ultimately evolve itself: Procyon, a subgiant, and Procyon B are good examples (Section 7.10).

If they are sufficiently close, however, tides will play a role. A tide is produced by a differential gravitational force across a body. The near side of the Earth is pulled harder by the Moon than the center, and the center harder than the far side. The result is a bulge on the surface of our ocean-covered planet that makes the sea rise and fall as the world turns. Tides occur in the solid Earth and Moon as well, and they all generate heat that slows the rotation rate. The end result is that orbiting bodies will synchronize their spins and orbits and keep one face always pointed toward the other. The mechanism is responsible for the Moon always facing the Earth, for the spin-ups of the RS CVn stars (Section 6.11) and the spin-downs of the Am stars (Section 8.3).

If the stars are not too near to one another, each will be distorted into a symmetrical oval shape, with the ends facing each other. If we could push the stars closer together, the distortion would increase, until the larger one reached a critical surface (called the *Roche lobe*, Figure 11.10), from which matter is lost, i.e. the star

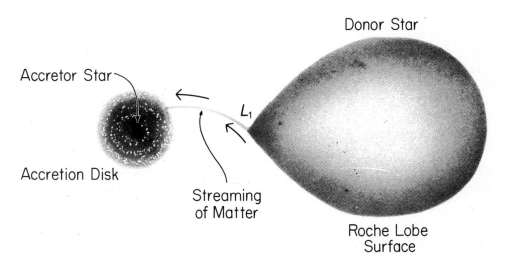

Figure 11.10. Mass transfer in a binary system. The small component of the binary, the accretor, produces a tidal Roche lobe surface around the larger donor star. If the donor can be forced past the critical surface, either by expansion during evolution or through increasing proximity (which decreases the size of the Roche surface), mass will be lost to the accretor through the 'Lagrangian point', L_1. The matter flows first into a revolving accretion disk and then onto the star. Many variations are possible. Donor/accretor pairs might be main sequence/white dwarf (nova, dwarf nova, Sections 11.4, 11.6); main sequence or giant/white dwarf (recurrent nova, Section 11.6); giant/main sequence or white dwarf (symbiotic, Section 11.8); giant or main sequence/black hole (X-ray source, Section 11.9). Adapted from *The Evolving Universe*, 2nd edn., by D. Goldsmith Benjamin/Cummings Publ. Co., 1985.

begins to come apart. As the first giant swells in a close binary it can encroach on its Roche lobe; matter will be directed outward at the point labeled L_1 in Figure 11.10 (the so-called *Lagrangian point*), whereupon it will flow to the dwarf companion, and the stars will begin to exchange mass. Since the giant can be enriched in heavy elements (witness the S stars and others, Section 4.7) the transfer can alter the atmospheric composition of the unevolved dwarf. The Sirius system provides a fine example. Sirius B, now the dim white dwarf, must at one time have been the more massive of the pair, and while it was a giant passed some enriched matter onto Sirius A, rendering it a mild Am star. We think the process may also be responsible for producing the barium stars (Section 5.5).

Now how does this picture relate to classical novae? If after the formation of the first white dwarf it and the main sequence star can be forced sufficiently close, the resulting tide will make the latter fill *it's* Roche lobe, with the result that it will now pass mass back in the other direction onto the surface of the degenerate. The fresh, accumulating hydrogen is compressed and heated by the intense gravity of the smaller star. Eventually we reach a critical point at which thermonuclear fusion commences and the surface of the star violently explodes, expelling the newly laid layer into space at high velocity. If the binary system is not too distant from us, or is not hidden behind clouds of obscuring dust, we might then be privileged to see a 'new star' burst into the nighttime sky, and slowly fade as the erupting gases blow outward into space. What we photograph before or after the actual outburst is never the star that produced the explosion, but the one on the main sequence that supplied the material.

How the stars get close enough is a matter of some debate. One possibility is that if they are born close enough together the first to evolve might actually encroach upon and encompass the other, with the result that both find themselves orbiting within a common envelope. Friction then causes a loss of energy, which makes the stars spiral even closer, so that after the first star becomes a white dwarf the two are synchronous rotators. Magnetic braking of the main sequence star of the type described in Section 6.5 then releases more energy, and because of the rotational locking the pair spiral together. The white dwarf raises tides in the unevolved member, which fills its Roche lobe, and the mass transfer starts, initiating the process that leads to the great blast. Theoretical certainty yet eludes us, however.

11.5 Nova remnants

We might expect that the expanding cloud of gas would become visible after it becomes large enough, and indeed it does. After a few years, depending upon distance and expansion velocity, a tiny shell is seen surrounding the perpetrating star (Figure 11.11). At this point, the developing nebula has a spectrum and an appearance quite similar to that of a planetary. But the likeness is cosmetic, as the nova shell has a much lower mass, is expanding 100 times faster, and has a visible lifetime of but a few centuries or so.

The shell provides us with a means of estimating the nova's distance. We easily know the expansion rate from the Doppler shifts in the spectrum. If we watch the

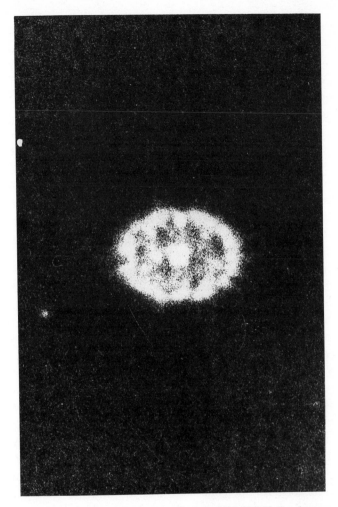

Figure 11.11. The expanding nebula around Nova Herculis 1934 (DQ Her), taken some 40 years after the outburst, by which time it had grown to a radius of about 0.02 parsecs. The star that supplied the fresh mass for the outburst is at the center; the white dwarf that actually produced the nova is now back to normal and is much too faint to be detectable. The system is an eclipsing binary (Section 5.4) that exhibits a short period 'flicker' caused by gas flowing to the white dwarf. This evidence was instrumental in the explanation of the nova phenomenon. University of Arizona photograph from an article by R. E. Williams, N. J. Woolf, E. K. Hege, R. L. Moore, and D. A. Kopriva in the *Astrophysical Journal*.

remnant grow for a few years we can also measure its angular expansion rate in seconds of arc per year. This quantity is like proper motion (Section 1.9), which depends on transverse velocity and distance. We assume that the expansion rate is uniform, and use the Doppler velocity to deduce distance. The method has been tried on the shells of the planetary nebulae but their growth rates are so much smaller that it does not work very well.

Chemical analyses of the shells suggests that they can contain large quantities of helium, nitrogen, and neon that were created in the explosion. If these excess abundances are real, and the point is still debatable, they may eventually tell us something of the nuclear processes that created the detonation. But as yet, we are unable to interpret the clues properly.

11.6 Dwarf and recurrent novae

Tied by name to the classical novae, but powered by a somewhat different mechanism, are the dwarf novae, also referred to as SS Cygni, U Geminorium, or Z Camelopardalis stars. These exhibit eruptions that take place with intervals that range from several days to a year. The shorter-period Z Cam stars will typically brighten suddenly by two to six magnitudes every 10 to 50 days; those of the U Gem variety will go off every 15 to 500 days or so. A Z Cam variable will also undergo an occasional hiatus in which the eruptions will be suspended for weeks to years at an intermediate brightness level. Superimposed on the eruption cycles of these stars are rapid flickerings that have periods of only seconds to minutes.

These eruptive variables are also caused by mass transfer from a main sequence star to a white dwarf. However this time thermonuclear detonations are not involved. The matter from the donor star in Figure 11.10 does not fall directly onto the receiving component, but first spirals around it, building into an *accretion disk*, from which it subsequently flows. A dwarf nova is powered by erratic changes in the disk as it stores and releases energy, and/or as it irregularly dumps matter onto the degenerate's surface.

There is a growing sentiment that the dwarf and classical versions are intimately connected with one another. As the main sequence component first overflows its Roche lobe (refer to Section 11.4) the transfer rate into the accretion disk is low, which causes it to be unstable and to behave erratically. We then see a dwarf nova. But as the two spiral together, the rate increases, the disk stabilizes, the irregular eruptions disappear, and the white dwarf begins to build a significant layer of hydrogen. Eventually it explodes, which drives the two stars a bit farther apart. The mass exchange halts for awhile, and as the stars move in on one another again, the process begins anew. It is a satisfying idea, one which consolidates two kinds of astronomical objects, but one that remains to be proven.

The sparsely populated set of recurrent novae may be related to both the classical and dwarf versions. These CV stars have been seen to explode more than once. The best known examples are T Coronae Borealis, which has reached second magnitude, with events in 1866 and 1946, and RS Ophiuchi, which can reach magnitude four and seems to go off every few decades. These two are apparently akin to the dwarf novae, in that the events are caused by instabilities in the accretion disks, but in these instances the donor is a red giant, and the matter falls onto a main sequence star. T Pyxidis and U Scorpii, however, seem to be thermonuclear detonations on white dwarf surfaces. Thus these provide some evidence that classical novae are recurrent too, but with long intervals between outbursts, that is we have only yet witnessed but one event.

It is exceedingly difficult to categorize the spectra of all these types of stars. The cataclysmic variables in general have some of the most complex spectra known, exhibiting features associated with an irregular gas flow, with the heating as the gas falls onto one component, and with sudden eruptions.

As a curious coda to this section, we might note that as disparate as the novae and planetary nuclei appear to be, there is an interesting parallel to their development. The steady contraction that accompanies the evolution of the pre-white dwarf nucleus of the planetary appears to be nearly identical to that following the nova outburst as the white dwarf settles back to normal, the difference being that the latter happens 100 times faster. Thus, our two left-over spectral classes, P and Q, whose natures were so poorly understood for decades after these letters were introduced, turn out to complement one another. The study of either can lead to new developments in our knowledge of the other.

11.7 Eta Carinae

This star, deep in the southern hemisphere, enmeshed in the great Carina Nebula, is odd enough to warrant its own section. We first encountered it briefly among the F stars (Section 7.10) where it resided for a time along with other strange luminous windy variables. For twenty years in the mid-nineteenth century it dominated its part of the sky, and in 1844 reached a magnitude of nearly -1, rivaling Sirius. In 1856 η Car began a slow 14 year decline to 7th magnitude. It underwent a mild one magnitude eruption around 1886, dimmed again, and since roughly 1948 it has steadily brightened to just about the naked eye limit. For many years this star was thought to be an extremely slow nova of the RT Serpentis class. The best known example of these is RR Telescopii, which burst forth in 1944, stayed near maximum for four years, and has not yet returned to its original state. But that classification almost certainly is not correct.

This extraordinary star is intimately involved with the great star-forming nebulosities in Carina (NGC 3324, NGC 3372; Figure 11.12) that contain numerous O stars, including the brilliant O3 Iaf supergiant HD 93129A that we looked at in Chapter 10. It is surrounded by its own small bright peculiar nebula (called the 'homunculus', as it is shaped like a tiny man.) The star itself is further buried and hidden within an even smaller dusty cloud of its own making and all we currently see in its optical spectrum are nebular emission lines. But old spectrograms taken in 1893 shortly after the small outburst show absorption features similar to those of an early-F supergiant (Figure 11.13). We believe that the star went through a major mass-loss episode and that the absorption spectrum was produced by a cool expanding veil ejected by the mysterious object within.

The unseen star is not understood. Our best guess is that it is a massive, very luminous O supergiant with an extraordinary absolute bolometric magnitude near -12.5. A binary companion may be involved in the outbursts; we do not know. Some astronomers think that it may be turning itself into a Wolf–Rayet star, while others speculate that it is a very likely candidate for an eventual supernova. Time, obviously, will either prove or disprove these notions, though the pace of stellar evolution may produce a long wait in human terms.

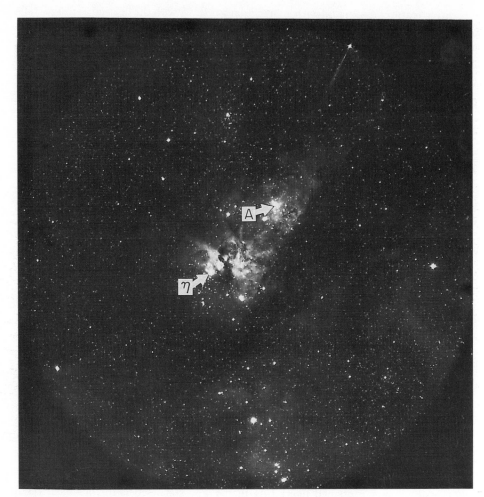

Figure 11.12. The Carina Nebula, NGC 3324 and NGC 3372. The weird variable η Carinae and the extraordinarily luminous O3 Iaf supergiant HD 93129A are marked 'η' and 'A' respectively. The dark lane to the right of η Car is called 'the Keyhole'. The nebula's prominence and context in the southern Milky Way can be seen in Figure 1.7, where it appears as a bright spot just to the right of the southern cross near the edge of the photo. National Optical Astronomy Observatories (CTIO) photograph, courtesy of D. A. Hunter and J. S. Gallagher.

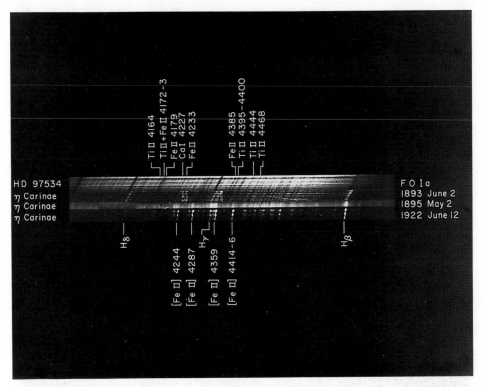

Figure 11.13. Spectra of η Carinae compared to that of the F0 supergiant HD 97534. We easily see the conversion of the absorption spectrum to one in emission as the outburst fades. The 1893 spectrum is one of the first of those taken at Harvard's Boyden station in Peru, which were used to classify the spectra of southern stars. From an article in the *Astrophysical Journal* by N. R. Walborn and M. H. Liller.

11.8 Combination spectra

The spectrum of an unresolved double star will, clearly, be a mixture of the two individual types. A prominent example is Capella, which exhibits a blend of classes G0 and G5. Similar examples abound, and include such stars as double-line spectroscopic binaries (Mizar, ζ UMa, for example, Figure 5.10) for which the individual sets of absorption spectra are separated by orbital Doppler shifts, and which are used to derive information on stellar masses. These are often called *composite spectra*, and are *not* the subject of this discussion.

True *combination spectra* are those that display syntheses of high and low temperature characteristics, and the stars that possess them are frequently termed *symbiotics*. A typical example exhibits an M giant spectrum and is an irregular variable with a period of perhaps a year or so, and an amplitude of about a magnitude. Superimposed upon this cool spectral background are common (permitted) emission lines of hydrogen, helium, and iron, plus forbidden features (Section 10.7) such as

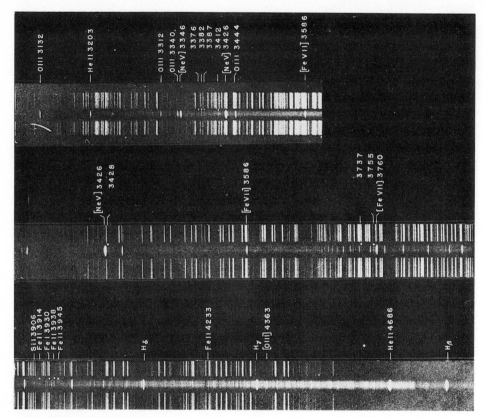

Figure 11.14. A combination spectrum, that of the symbiotic star CI Cygni. The type M background shows TiO bands. High excitation features such as He II and [O III], which can only be caused by a hot source that ionizes gas in the neighborhood of the two binary components, are superimposed upon it. McDonald Observatory spectrogram, from an article by J. Sahade in *Stellar Atmospheres*, J. L. Greenstein, ed., University of Chicago Press, Chicago, 1960.

[O III] and [Ne III]: see Figure 11.14. At shorter optical wavelengths, we see the blue continuous spectrum of a hot source superimposed upon, and overriding, the M star background. The far ultraviolet, as observed by satellite, is especially impressive, with numerous high-excitation emission lines. P Cygni features may be present, indicating an outflowing wind, and the far infrared may show radiation from a cocoon of surrounding dust. The hot source and the emission features are quite highly variable, but are more or less independent of the behavior of the M star. The hot blue continuum exhibits occasional outbursts every few years or decades, and might dominate a good portion of the optical spectrum for a year or more before returning to its normal state.

We see wide variations upon this summary theme. The cool component may be class K or even possibly G. In some, the hot source and the emission lines are barely detectable; in others they overwhelm the late-type star, and we know that it is present only from a few weak TiO absorption features. The temperature that characterizes the blue continuum may lie within a very broad range, and the normally-found high-excitation He II lines may be weak or even absent, or they may dominate the emission spectrum. The forbidden lines, which can be suppressed by high density, may or may not be present.

Good and quite bright examples of symbiotic stars abound. The prototype is Z Andromedae, which at normal maximum is about eighth magnitude. Symbiotics are sometimes even referred to as Z And stars. The brightest is CH Cygni, which has been in a 'blue outburst' state for nearly twenty years, and which is easily visible in binoculars about 6° northwest of δ Cygni (it can be found on a good quality amateur's sky atlas). A more bizarre relative is R Aquarii, which is enshrouded by a large optically visible nebula of its own making.

An early idea based on the presence of the nebular emission lines held that the symbiotics might be in the process of producing new planetary nebulae. Other single-star theories involved energetic shock waves in an envelope expanding around a cool star or a low temperature mantle over a hot star. But it is now widely accepted that the key to the explanation of the phenomenon again lies in stellar duplicity.

The observations can be understood quite well by postulating that the cool giant is losing mass to a dwarf or white dwarf companion (again like the scene in Figure 11.10). The hot source may be the secondary star itself, but it is more likely that we are observing a hot spot on the star caused the infalling matter itself, or that we are seeing a hot accretion disk that radiates much as if it were a star. In either instance, light variations can result from changes in the rate of mass loss from the giant. The hot spot in turn can ionize a large volume of the infalling gas, which then radiates the nebular lines and can contribute significantly to the blue and ultraviolet continuum. If the gas density is low, we will see recombination lines of hydrogen and planetary nebula-like forbidden features. If the density in the flow is high, the forbidden lines will be suppressed as the electrons that create them will be removed from their upper levels by collisions. The temperature, or activity level, of the hot spot will control the degree of ionization, so that for some stars we will see the lines of ionized helium (e.g. λ4686 of He II), for others only neutral helium.

Since all this free matter is in orbit, it can be periodically eclipsed by the giant, resulting in regular variations superimposed upon the irregularities caused by fluctuations in the mass flow. Occasionally, the accreted matter may ignite in a nuclear-fusion reaction, which causes the symbiotic to brighten very suddenly. It can stay in this 'high state' for months, or like CH Cygni, for many years. Symbiotics then can be explained much like novae: mass transfer is the key. The ultimate explanation in both cases, and for the planetary nebula phenomenon as well, is stellar evolution. We will consider that topic in detail in the next and final chapter.

11.9 Beyond the HR diagram

We have nearly completed our survey. We have covered an immense range on the HR diagram, from under 2000 K to over 50 000 K, from stars 10 000 times fainter than the Sun to those nearly 10 million times brighter. If we modify and expand the diagram to a plot of true luminosity against temperature, as is done for theoretical evolutionary studies, we can also include the planetary nuclei, and extend the limit to well over 100 000 K. Within these confines, we have seen many different kinds of stars, including the variable and the chemically odd. We have not looked at all of the varieties, but enough to have captured the flavor of modern stellar astronomy, and to enable the reader to pursue the subject further.

But there are yet two classes of stars to call attention to that cannot be encompassed by any change in the form of the HR diagram, and which are unclassifiable under the standard systems. These are the extreme end products of massive star evolution: the neutron star (or pulsar), and the black hole. The spectrum of the first is continuous and has no absorption or emission lines, since densities are so high that there are no atoms as we know them. The second not only has no lines, it has no light, as gravity prevents the escape of radiation. The neutron star is the next stage beyond the white dwarf, and must have a mass greater than 1.4 times solar (Section 8.9). However, like its smaller cousin, it too has an upper mass limit, and will collapse if greater than about 4 times that of the Sun. The stellar black hole, whose collapse cannot be stopped at all, is the next step and is expected to exceed this limit; beyond it there seem to be no bounds. We will go into these matters in more detail in Chapter 12.

We can identify stellar black holes (as opposed to much more massive ones that may reside at the cores of galaxies) only when they are components of binary systems. We might, for example, be able to infer a sufficiently large mass of an invisible object from the orbital deflection that it gives to a visible companion, much as we might do for brown dwarfs (Section 4.10). If such a strange body is involved in a mass-transfer type binary (Figure 11.10) we may see its effect more directly. As matter from the normal star falls into the black hole it is heated and compressed so that it radiates fiercely, even producing X-rays, shortly before it disappears forever. Candidate X-ray stars are indeed observed. Both the neutron star and the black hole are by-products of super-novae: the extraordinary explosions that terminate the lives of the most massive stars.

Now our stellar inventory ends and we are ready to examine the reasons for this superb variety presented to us by the heavens.

12 *Journeys on the HR diagram*

On an April evening, Arcturus ascends the eastern sky, and as we admire its orange color we might contemplate its role as a superb example of a K star. We identify it as such, much as we would categorize ourselves as woman or man, or our pet as dog, cat, or armadillo. But it may be more meaningful, perhaps, to classify instead in terms of periods of maturity: child, adult, aged. Then we might think to question whether Arcturus always looked as it does today, and ponder how it may have appeared in the distant past, and what the future holds for it.

For the stars must always be changing. Their internal engines run on a finite amount of fuel, which means that their lifetimes, though terribly long by our standards, are limited. Interior changes caused by the deterioration of the energy supply produce profound exterior alterations that will result in the movement of a star on the HR diagram. Arcturus may once have been an A star, and might at some time appear as a class M giant.

What began above as a simple speculation has in this century been turned into a major part of astrophysics: the subject of stellar birth, evolution, and death. After an immense effort and countless hours of calculation, we have arrived at the astounding conclusion that almost all stars have passed or will pass through all or most of the major spectral classes, and that the great variety of celestial objects can be tied together by an understanding of the aging process. We sketched the barest of outlines of this subject in Section 1.17, and have alluded to it several times in various passages. Now, armed with the descriptions of the wonderful array of stars covered in previous chapters, we can look at it in close detail.

12.1 Concepts

Accurate, detailed calculations of evolutionary developments can be enormously difficult to perform, and in many instances we do not know how to make them at all. But the underlying principles are remarkably simple and straightforward. The ultimate cause of stellar change is the constant attempt of a star to contract under the steady pull of its own gravity. All the various stages that we identify are produced either by processes that temporarily halt the contraction, or by the act of compression itself.

The most important factor – to a good approximation the only factor – is the star's initial mass. We can think of a star on the main sequence as having three rather distinct layers (refer to Section 1.16): a nuclear burning core in which the stellar energy is generated; a thick envelope, which blankets the core and helps maintain its temperature and pressure and which transmits the energy to the outside; and the thin outer atmosphere in which the characteristic stellar spectrum is formed. Gravitational compression by the whole star is responsible for heating the core to temperatures above the nuclear flash point. Higher mass stars will have greater internal temperatures and concomitantly larger core masses or fuel supplies. More importantly the available fuel, being hotter, will produce energy at an increased rate per unit mass.

As a consequence of these effects higher mass stars are by far the more luminous. We do not need theory to tell us this fact, though. From measurements made of binary star systems (Sections 1.6 and 5.4) we establish the empirical mass–luminosity law, which in its simplest form shows us that along the main sequence the luminosity, L, is on the average proportional to mass, M, to roughly the 3.5 power (Section 7.2). For a variety of reasons outlined briefly below, the exponent actually changes considerably as we progress upward in mass from the dwarf M stars; the above value should be used only as a general guide. The triumph of the theory is that it can reproduce the true relationship almost exactly. It is largely from this agreement with observation that we can ' prove' our contentions as to what makes the stars work.

If we were to attempt an instant judgement, we might think that the most massive stars, having the most fuel available, should live the longest. But as pointed out on several previous occasions, the true case is exactly the opposite. Stellar duration depends upon the amount of matter that can be burned divided by the rate at which it is used. Consumption is so rapid for high mass stars that it easily offsets the greater quantity of useful matter, which produces dramatically shorter lives. Let us use simple, average properties of stars to see just how much life expectancies do change. The approximate lifetime of a star, t, will be roughly proportional to its mass M divided by its burning rate, which in turn is proportional to the luminosity L, so $t \propto M/L$. But as we saw above, L is on the whole proportional to $M^{3.5}$. Consequently, t must scale as $M/M^{3.5}$ or $1/M^{2.5}$. From our knowledge of solar structure, our Sun initially had enough fuel to enable it to live for 10^{10} years. A ten solar mass star will then endure for only $1/10^{2.5}$ or 1/300 as long, or only 30 million years. Toward the other extreme, a star only 1/10 as massive as the Sun will survive for 3×10^{12} years before it succumbs.

These numbers are meant only to be illustrative and can easily be a factor of two or more in error. To produce the mass–luminosity–lifetime relations accurately we need to include many complicating factors: the inconstant proportionality along the main sequence between total mass and that actually available for burning (that is, the ratio of core and envelope masses); variations in initial chemical composition that can alter the rates of the generation of energy and its transfer rate through the envelope; the enlargement of the core that takes place during evolution; convection that can add fresh fuel to the core from the envelope; and mass loss, which we know is so prevalent, and which can change the interior construction. It is out of these intricate details that the science of stellar evolution is created.

The above arguments are concerned primarily with the conversion of hydrogen to helium, which takes place on the main sequence (see Displays 6.1 and 7.1). At the next level of complexity above our initial discussion, we must consider all the various stages of nuclear burning, which directs our attention to stars that lie elsewhere on the HR diagram. Just as a star in its contraction attempts to achieve the lowest energy configuration by making itself as small as possible, so do the fusing atoms of the stellar core. A star performs this feat by becoming a white dwarf, a neutron star, or a black hole; an atomic nucleus does it by becoming iron. To create energy through nuclear fission, we use heavy atoms: uranium, radium, plutonium; generation by fusion employs light atoms: hydrogen, helium, carbon. The component parts, the protons and neutrons, of iron are more tightly bound than in any other atom, and therefore the transmutation of this element by either fission or fusion requires an *input* of energy. This is the reason why iron is by far the most common of metals.

The many stages of a star's life after the main sequence can be understood in terms of the material that is being fused, or is about to be fused, at a given time. Light elements fuse sequentially into heavier ones as the temperature and density of a nuclear burning core increase with age and compression. Hydrogen first burns to helium. That ash of that first nuclear fire then ignites to form carbon; C later can go to neon, neon to silicon, and Si to Fe. How far the process develops depends again upon the initial stellar mass and the complicating factors discussed above. With this outline as our background, let us now look at the intervals of a star's life, and connect the periods of contractions, pauses, and nuclear fusion stages with what we see in the nighttime sky.

12.2 Star birth

The first great contraction stage creates the star out of the chaos of the interstellar medium. We are now beginning to see how it happens; we cannot yet follow the progress of a star through its entire birthing sequence, but we do see some marvelously intriguing clues to what happens, and we can recognize some nascent stars in the process of their formation.

It all must begin with condensations deep inside cold clouds of gas and dust (Figure 12.1), much like Barnard 86 pictured in Figure 9.11. We know that massive young stars are very frequently found buried within these complexes. Sometimes the clouds are so thick and opaque that the new associations and clusters cannot be seen in the optical part of the spectrum, and we must rely on infrared detection in order to penetrate 10 to 20 magnitudes of visual extinction (Figure 12.2). Also found associated with dusty nebulae are the young T Tauri stars (Section 5.6 and below) that lie above the main sequence and appear to be contracting toward it.

The actual formation and development of the condensations seem to involve the interplay between gravity and the tangled magnetic fields that pervade the cloud. The major problem in star formation has historically been the difficulty of angular momentum, specifically how to get rid of it. Angular momentum is a measure of the amount of rotational or orbital energy in a body or system of bodies. If you whirl a rock on a string around your head, its angular momentum, called L (do not confuse it with luminosity), is equal to the rock's mass (m) times its velocity (v) times the length of the string (r), or

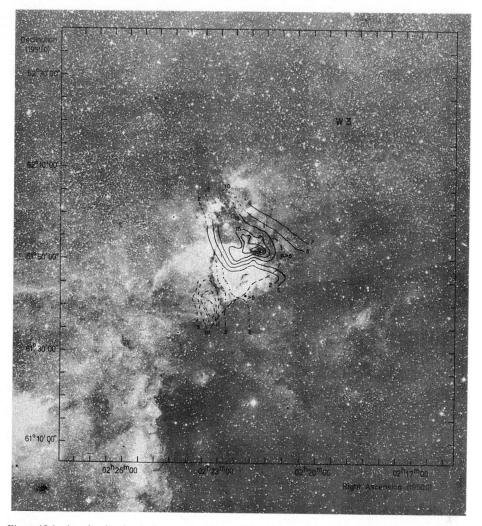

Figure 12.1. A molecular cloud, where stars are born. The contours, derived from radio spectral line studies, overlay a complex called W3 and show the concentration of carbon monoxide. Star formation can begin within these cold zones, where the temperature may drop to only a few degrees above absolute zero. Photograph © National Geographic–Palomar Observatory Sky Survey, reproduced by permission of the California Institute of Technology. Radio contours superimposed by H. R. Dickel.

$L = mvr$. In a spinning body, L is the sum of the angular momenta of all the parts (think of them as individual cubic centimeters) as they turn about the axis.

Like energy, L is conserved, that is in a closed system with no outside influences, it must stay constant: if r is decreased, v increases in response. This rule explains why a spinning skater speeds up when his or her arms are brought in and why the Moon is moving away from the Earth as lunar tides slow our planet down.

Interstellar clouds do not spin very rapidly, but they are huge, and consequently contain a vast amount of angular momentum. As one shrinks its rotation speed must go

Figure 12.2. Buried stars. The arrows mark positions of young B and A stars immersed in dusty clouds. They suffer over 10 magnitudes of absorption in the visual, and can be seen only in the infrared. Photograph taken by G. Grasdalen with the Curtis Schmidt telescope at Cerro Tololo Inter-American Observatory, from an article in the *Astrophysical Journal* by K. M. Strom, S. E. Strom, L. Carrasco, and F. J. Vrba.

up, and one can easily demonstrate that the velocity becomes so great that a nascent star would be torn apart long before it could even begin to approach main sequence status. As a cloud compresses under gravity it fragments to produce binaries as well as single stars, with the collection ultimately becoming a galactic cluster. A large amount of spin angular momentum can be translated into orbital motion, but even this process falls far short of the amount needed to be removed.

It is here that magnetic fields come into the picture. As the cloud's embedded fragments – the stars and doubles to be – contract, they carry their entrapped magnetic fields with them. The field lines act like ropes with one end caught within the budding stars and the other end tangled in the remaining gas of the cloud. These lines then act as brakes that slow the spinning fragments. At a certain point, the conditions within the fragments are such that they can no longer hold the ropes, the brake is released, and we are left with a rotating blob of gas that has an angular momentum below, or at least near, the critical limit that would allow formation. If needed, separation into a binary system can now finish the job. Although this picture appears superficially complete, there are many gaps and uncertainties; keep in mind that in this difficult field much of what we think we know may be wrong.

12.3 Developing stars

We actually find quite a variety of stars in various stages of development and can string them together so that we at least begin to see how these collapsing blobs of gas are converted into main sequence dwarfs. Figure 12.3 shows a dense, cold star-forming region between σ Scorpii and ϱ Ophiuchi. A wider scale photo can be found in Figure 4.1, which shows all of northern Scorpio, with the distinctive ϱ Oph nebula near top center (it is quite interesting to compare the detail in the two fields, one obtained with a small Ross camera, the other with the 1.2-meter United Kingdom Schmidt telescope). The cloud is also seen at the right of Figure 9.1, where it is well shown within the context of the extended dusty regions of Ophiuchus. The inset (not to the same scale) shows a newly-formed cluster, not unlike the region depicted in Figure 12.2, buried so deeply within the ϱ Oph dark cloud to the south of that star that it can be detected only in the infrared. To the left within the inset is a remarkable infrared object known as IRAS 16293-2422. (IRAS stands for Infrared Astronomical Satellite, an orbiting telescope that mapped the sky in four infrared bands during 1983; the numbers refer to the coordinates of the source, where the first number gives the right ascension and the second the declination: see Section 1.3). Observations of it in a radio emission line produced by carbon monosulfide (CS) clearly show infalling gas: apparently we are seeing a star in the actual act of being created. Estimates of its mass (only one-quarter solar) and of the accretion rate suggest that the object is only about 30 000 years old.

Further remarkable clues to the process of star formation come from the *Herbig–Haro objects* and the *bi-polar nebulae*, which are loosely connected. The former are small bits of bright gaseous matter (named after G. H. Herbig and G. Haro) that appear to have no central illuminating sources as would planetary nebulae. An example is shown in Figure 12.4. They are almost always found in the neighborhood of the

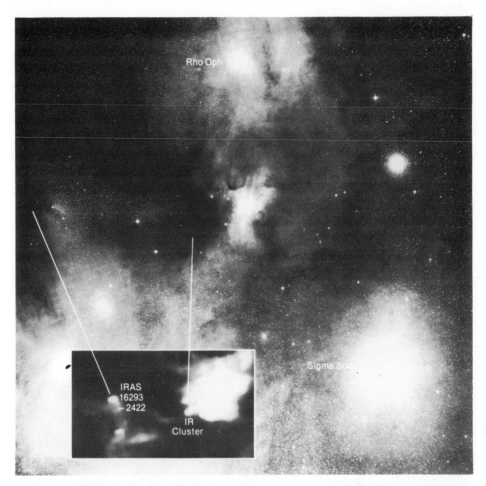

Figure 12.3. Birth of a star. The large optical photograph shows the bright and dark nebulae in the neighborhood of ϱ Ophiuchi, a binary that consists of a B2 dwarf and subgiant; see Figure 4.1 for a wider view. The inset displays infrared images of a cluster within the dark cloud south of ϱ Oph, and of a star, IRAS (Infrared Astronomical Satellite) 16293–2422 (the numbers refer to right ascension and declination), that appears to be in the process of formation. Main figure: UK Schmidt Telescope, copyright Royal Observatory, Edinburgh; inset from the Infrared Astronomical Satellite processing and analysis center; panel from an article in *Sky and Telescope* by C. J. Lada.

young T Tauri stars (Section 5.6), and they exhibit an intricate structure that changes with time. Proper motion studies of some groups of them show that they are moving radially away from a central source of energy. Apparently, newly-formed or developing stars are beaming matter outward in opposite directions. This high-speed gas, whose conditions – temperature and density – are unknown, rams into ambient gaseous clumps, heats them by collisional shock, and then accelerates them away. It is not known to what degree the mass ejected from the exciting star may actually add to and supplement the Herbig–Haro blobs themselves.

Figure 12.4. Herbig–Haro objects in Orion. The bright blobs are moving radially outward from an optically invisible star and are propelled by an interaction with a strong wind. The arrows point to the location of a bright source seen only in the radio spectrum and in the long-wave infrared that has a luminosity of about 50 times that of the Sun, and a mass of about 20 solar. We cannot see the object because it is buried in a dust cloud that produces some 50 visual magnitudes of absorption. Lick Observatory photograph, from an article in the *Astronomical Journal* by G. H. Herbig and B. F. Jones.

Figure 12.5. R Monocerotis, at the lower tip of NGC 2261 – 'Hubble's Variable Nebula' – is one end of a bi-polar nebula; the counter-fan, below R Mon, is sometimes barely visible, and appears to be heavily obscured by dust. Both components are visible in radio observations of carbon monoxide, a molecular tracer of intermixed dense, cold gas. Photo taken by E. Hubble with the 100-inch Mt. Wilson reflector. Mt. Wilson and Las Campanas Observatories, Carnegie Institution of Washington, from an article by A. H. Joy in the *Astrophysical Journal*.

Our ideas are reinforced by the phenomena of the bi-polar nebulae (Figure 12.5). In these, we can actually see two fan-shaped jets emerging from opposite sides of a central location. The apparent source of the gas sometimes contains an observable star, but occasionally the dust obscuration is so heavy that nothing can be seen optically. In order to examine the central generator of energy we can, however, employ radio or infrared techniques that penetrate this shroud. The FU Orionis stars (Section 7.10) are all surrounded by dust and odd reflection nebulae, and are probably extreme members of the broad class of bi-polar objects. In addition to the spectral signature of infalling matter, IRAS 16293–2422 (Figure 12.3) also shows evidence for *outflowing* gas, demonstrating that it too has the character of a bi-polar object, thereby tying the various protostar candidates together. In ways not yet understood, mass accretion drives mass loss.

It seems from all the evidence that as a star forms, the remaining material flattens out into a thick rotating disk: matter is then shot out through the poles by way of an uncertain acceleration mechanism that may involve rotation and magnetic fields, thus creating the jets of the bi-polar nebulae, or interacting with interstellar gas to form the Herbig–Haro objects. We do not yet know enough to be able to string the various kinds of objects together with real confidence into an evolutionary sequence. But it is exciting to think that the disks we see, or infer, could be the breeding zones for the stars' families of planets. The remains of these disks are also now being found: examples are the cloud

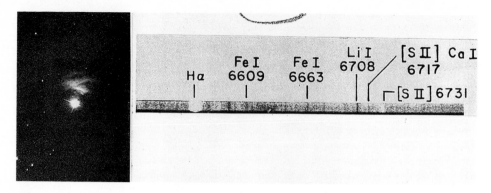

Figure 12.6. T Tauri and its spectrum. This extraordinary young variable has recently contracted, and is seen embedded in the remnant gas. The spectrum exhibits bright lines of hydrogen and forbidden sulfur. The emissions can sometimes take on P Cygni features, indicating an outflowing wind, and remarkably, occasionally show *inverse* P Cyg lines that characterize *infalling* gas. Extreme youth is proven by the abundance of lithium: from the line at $\lambda6708$, we find 100 times as much as we see in the Sun. This element is destroyed by nuclear processes in mature main sequence stars, but T Tau still has the Li composition of the interstellar medium. From the *Astrophysical Journal*: Photograph from Mt. Wilson and Las Campanas Observatories, Carnegie Institution of Washington, by W. Baade with the 100-inch telescope; spectrogram, Palomar Observatory, Cal. Inst. of Tech, by W. K. Bonsack and J. L. Greenstein.

discovered around Vega with the infrared satellite IRAS and the flat structure that extends outward from the A5 dwarf β Pictoris (Section 8.6 and Figure 8.9). And perhaps we should even include the disk or plane that contains our own solar system.

Theoretically, a star will become visible during its contraction phase as it enters the HR diagram from the right. At some point, the temperature in the core becomes hot enough to sustain thermonuclear reactions, and the star will settle into its initial position on the *zero-age main sequence* (the *ZAMS*) with a full hydrogen fuel supply. Shortly before final stabilization it passes through the T Tauri phase. Further evidence for the youth of these stars – apart from their connection with interstellar matter – is their lithium abundance. Recall from Section 6.10 that this element is destroyed in solar-type stars as they age. The T Tauri stars have the full interstellar complement (Figure 12.6), 100 times that found in the Sun, consistent with their being very young.

The location of the ZAMS depends upon the chemical compositions of the stars being formed. The metallicity of newly born generations of stars continually increases as older stars evolve and enrich the birthing interstellar medium with heavy elements. The resulting increased opacity of the atmospheres displaces the ZAMS ever further to the right on the HR diagram. The subdwarfs of Population II arise from a ZAMS shifted to the left of the Population I (solar abundance) version that we consider here as standard.

We are still far from explaining many important phenomena and details. For example, we do not know how to reproduce the range of observed masses, or rather the luminosity function, which we examined in Section 10.9. We also do not yet really understand the natures of the tracks that stars of various masses take to arrive onto the main sequence. But our progress over the past two decades – before which time this

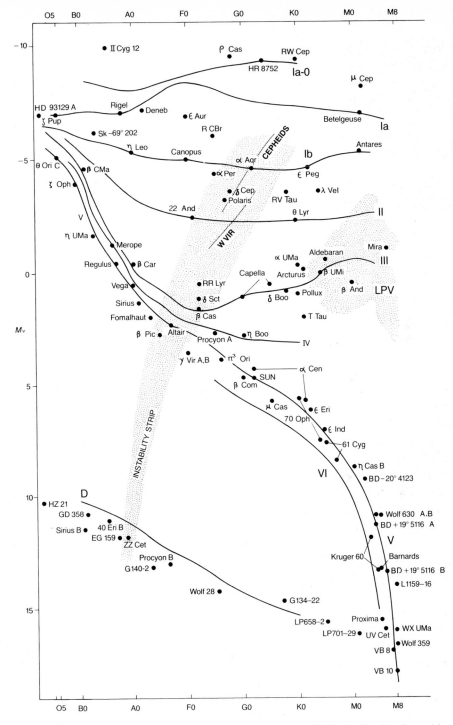

Figure 12.7. The completed HR diagram. Stellar data were taken from *The Bright Star Catalogue*, 4th rev. edn., by D. Hoffleit with the collaboration of C. Jaschek, Yale University Observatory, 1982, and from *Astrophysical Quantities*, 3rd edn., by R. H. Allen, Athlone Press, London, 1973. See Figure 3.8 for further credits.

subject was enshrouded in real mystery – has been remarkable. Fully detailed explanations of the processes involved should not be long in coming.

Now the stars are in place. We have spent the past eight chapters presenting them,
their characteristics, and their locations on the HR diagram. Now we combine all these
data to present this fundamental graph in its full form in Figure 12.7. The remainder of
this final chapter now serves to tie all these stars and their phenomena together and to
provide a theoretical understanding of why the various kinds exist.

12.4 On the main sequence

Formation represents the first step in the inexorable contraction of a star, and the
main sequence is the first hiatus. The great squeeze is halted for a long pause by
hydrogen burning. During this period, a star becomes a remarkably stable nuclear
furnace. We might logically anticipate that as the hydrogen is used up, the brightness of
the star should slowly decrease. But note here that the pressure of a gas, and its ability
to withstand a compression, depends only upon the number of atoms per unit volume,
not their kind. In the stellar core, four hydrogen nuclei are converted to one of helium
(Displays 6.1 and 7.1), which allows for a very slow contraction that also raises the
temperature. The resulting increased rate of nuclear burning very neatly offsets the
decrease in the amount of available fuel.

Changes do occur, but they are not large compared to what happens later. The net
effect over the dwarf lifetime is to cause a star like the Sun to become a bit brighter as
well as to expand and heat slightly at the surface. The result is that it begins to move
slowly up and to the left of its zero-age position as the fuel supply starts to diminish
(Figure 12.8). The Sun, for example, was at the time of its birth nearly five billion years
ago about 5% smaller and dimmer than now and some 200 K cooler. During the next
five billion, while the remainder of the core hydrogen is being consumed, these changes
will accelerate, and by the end of the period our star will be a very noticeable 25% larger,
200 K hotter and twice as bright as it was when it began its life on the ZAMS. These
comparatively small changes, coupled with a mix of starting metallic compositions,
account for the breadth of the observed main sequence of Figures 3.6 and 12.7. To a
crude approximation we can observationally track a later type star (G or K) through its
long dwarf lifetime by monitoring the decline of its lithium abundance (Sections 6.10

Figure 12.8. The early evolution of the Sun, demonstrating how luminosity and surface temperature change
with time. The three curves show the evolution of a one solar mass star on a graph with the logarithm of the
temperature plotted against that of the luminosity. Temperature and luminosity values are given on the
outside scales, the dwarf spectral classes are given across the top, and the absolute bolometric magnitudes are
on the right. The three curves show the effect of different helium mixtures: He/H = 0.107 for the thickest
line, which is closest to the Sun, 0.0625 for the thinnest, and 0.083 in between. The numbers show the time
(in billions of years) since birth on the zero-age main sequence. The dashed lines are loci of constant radius.
The evolution is slow up to about 10 billion years as the Sun moves through the band that defines the main
sequence, and then, with the exhaustion of internal hydrogen, rapidly climbs to become first a subgiant and
then a giant. The effect on Earth is profound. Note the sensitivity of the evolutionary timescale to helium
abundance, one of the many complicating factors in evolutionary calculations. Adapted from a figure
prepared by I. Iben, Jr. that appeared in *Annals of Physics*.

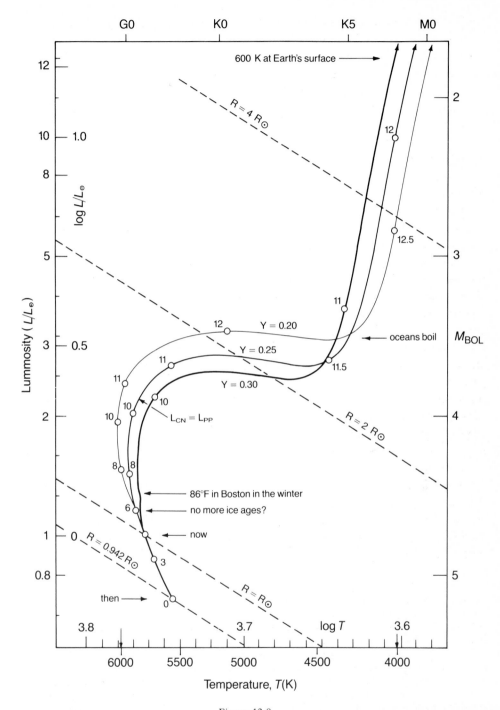

Figure 12.8.

and 12.3) and the decrease of its rotation speed and chromospheric activity level as the magnetic fields apply the brakes to its spin (Section 6.5). We owe our lives to the zone of main sequence stability: without it, life would never have had time to develop on Earth.

12.5 Giants and supergiants

The real fascination with stellar evolution begins as the internal hydrogen nears exhaustion. When the H abundance goes below about 1%, no adjustment in burning rate can keep up with the failing supply, the central fire effectively goes out, and the core begins to contract much more quickly. We might again intuitively expect the star to become dimmer, but with the exhaustion of the fuel and the demise of a critical mechanism for internal support, it can once again use its immense supply of gravitational energy. As the core slowly collapses, its temperature continues to increase, fresh hydrogen in the envelope ignites in a shell around its now-quiet helium constitution, and the combination of all the internal effects causes the star to become vastly larger, and in most cases much more luminous. The size increases so much that the surface temperature drops dramatically (in keeping with the rule that luminosity is proportional to surface area times the fourth power of the temperature: see Section 1.15), and we now see developing giants and supergiants as the stars move up and/or to the right on the HR diagram.

This subject is studied largely by means of the calculation of *evolutionary tracks* (Figures 12.8, 12.9, and 12.10), which follow the positions of stars on plots of the natural physical variables, luminosity and surface temperature (strictly speaking, the logarithm of the luminosity plotted against that of the temperature). Such a graph, usually called the 'log L–log T plane', is a common variant of the empirical HR diagram

---

Figure 12.9. Evolutionary tracks on the log L–log T plane for Population I stars of differential initial masses. Both T and log T are given in the lower axis; log L, on the left, is simply the exponent of luminosity (in solar units) in powers of 10. Spectral classes are displayed across the top with those appropriate to dwarf and supergiant temperatures above and below the line respectively. Absolute bolometric magnitude is given on the right. Remember the bolometric corrections to visual magnitudes, which are over one magnitude for stars later than K9 and earlier than B7 (see Table 12.1). The heavy dots on each track represent, in order, the zero-age main sequence, the onset of core contraction following hydrogen exhaustion, and the beginning of helium burning. The separation between the first two dots represents roughly the spread of the observed main sequence. *IS* is the instability strip of Figure 12.7. The Population I 'clump' (C), where core helium burning takes place, the analogous Population II horizontal branch (HB), and the second ascent following helium depletion for a Population I star (AGB), are indicated for a standard solar-type (or just sub-solar) case of a $0.55 \, M_\odot$ core. The fourth heavy dots on the 5 and 9 $M_\odot$ tracks show the second and major phase of core helium burning for more massive population I stars.

The planetary nebula is formed at the top of the AGB, the star continues to the left, and the nebula becomes visible at the mark noted 'PN'. This track then continues on to Figure 12.14 below, and after exiting that figure reappears here at the lower left, marking the path of a white dwarf (WD). The positions of the tracks can be shifted by changing the chemical compositions of the theoretical stars, and by altering the parameters that control calculations of such things as convection. The reader should mentally apply the 1 $M_\odot$ solar track to the HR diagram of Figure 12.7, to see what may become of the Sun. Adapted from an article by Icko Iben, Jr., reproduced with permission from the *Annual Review of Astronomy and Astrophysics*, vol. 5, 1967, by Annual Reviews, Inc.

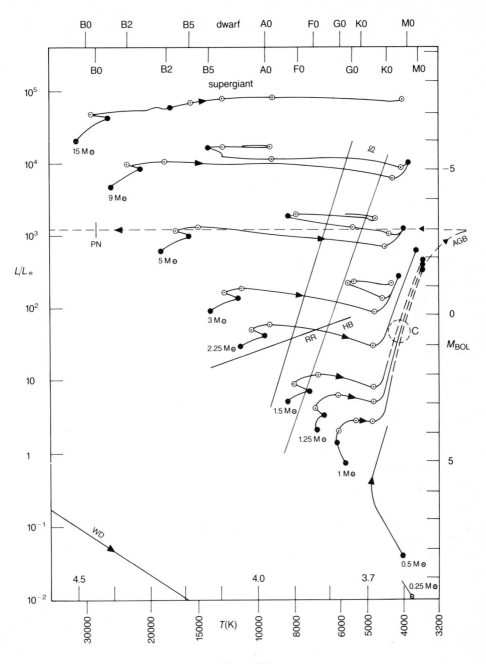

Figure 12.9.

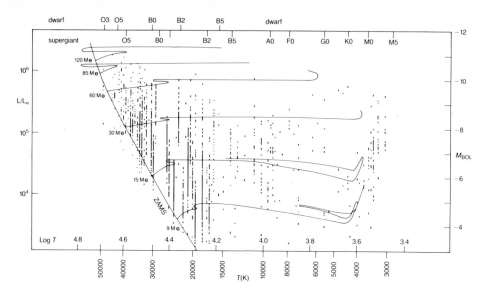

Figure 12.10. The distribution and evolution of massive stars, displayed in a fashion identical to that of Figure 12.9. Initial stellar masses, in solar units, are shown to the left of the zero-age main sequence (ZAMS). The evolutionary rates are at first relatively slow, producing a broadened main sequence. The stars then evolve horizontally as supergiants, and pass rapidly through the wide Hertzsprung gap in classes F, G, and K. Finally the rates diminish and the stars pile up as M supergiants. Adapted from an article in the *Astrophysical Journal* by R. M. Humphreys and D. B. McElroy; evolutionary tracks are by A. Maeder.

that we have used throughout this book. Conversions are readily made by employing bolometric corrections (Section 4.3 and the introduction to Chapter 10: see Table 12.1 for a summary) and careful observations of the temperatures of the stars in the various spectral classes (Section 3.7 and Figure 3.10).

The details of stellar evolution are enormously complex. The exact track that a star follows depends very strongly upon its initial mass, rate of mass loss, and chemical composition. A compendium of the complete collection of such paths for all stars would require a great deal of space and would add little to our discussion. Standard sets that take the stars into the giant and supergiant zones are presented here only to illustrate the basic idea of the results of the evolutionary processes at work.

The most important determining factor is the mass. From Figures 12.9 and 12.10 we see that lower mass stars behave quite differently from those of higher mass. A transition takes place between about three and nine solar masses (abbreviated $M_\odot$). Below that figure, the stars become more luminous as they evolve and move into the giant realm of the HR diagram of Figure 12.7. Near the top of the range and above it the evolutionary tracks flatten out and the stars become supergiants: around 10 $M_\odot$ for the Ib variety and 20 for the Ia. The hypergiants, class O at the top, absolute magnitude -9 and above, result from stars of 30 $M_\odot$ and up.

It is interesting to compare the evolutionary tracks where they overlap between Figures 12.9 and 12.10. Note that the two 9 solar mass paths, as well as each of the 15

Table 12.1. *Bolometric corrections*

The bolometric corrections are added algebraically to the visual magnitudes to derive bolometric magnitudes (Section 4.3). Numbers given to one decimal place should be considered approximate. The small positive values among the F giants and supergiants result from minor magnitude calibrations. These data were compiled from articles by R. M. Humphries and D. B. McElroy in the *Astrophysical Journal* and by D. M. Popper in *Annual Review of Astronomy and Astrophysics*.

Class	Main sequence	Giants	Supergiants
O3	−4.3	−4.2	−4.0
O5	−3.9	−3.8	−3.7
O7	−3.6	−3.4	−3.3
B0	−3.00	−2.9	−2.7
B1	−2.50	−2.00	−1.7
B2	−2.0	−1.6	−1.35
B3	−1.8	−1.5	−1.15
B5	−1.44	−1.3	−0.82
B7	−0.94	−1.07	−0.64
A0	−0.15	−0.24	−0.3
A5	−0.02	−0.02	0.00
F0	−0.01	0.01	0.14
F5	−0.03	−0.01	0.13
G0	−0.10	−0.13	−0.1
G5	−0.14	−0.34	−0.20
K0	−0.24	−0.42	−0.38
K5	−0.66	−1.19	−1.00
M0	−1.21	−1.28	−1.3
M2	−1.75	−1.52	−1.5
M4	−2.28		−2.50
M5	−2.59		−3.3
M8	−4.0		

$M_\odot$ loci, are noticeably different from one another: those in Figure 12.10 have lower luminosities, and the loops are of different lengths. The stellar computer models used in the two sets of calculations are different, with different input parameters, which illustrates at least some of the theoretical uncertainties. The chief difference is that the high mass models of Figure 12.10 include the effects of mass loss, whereas those of Figure 12.9 do not. The former are probably more realistic, although the mass loss rates, as well as the effect of such losses on the stars, are still not very clear to us.

However, even a simple, coarse perusal of the calculations, coupled with accurate computations of stellar lifetimes, can explain much of what we observe. To begin, the initial move off the main sequence is quite rapid, which results in a paucity of stars to the right of the broadened main sequence in the central portion of the HR diagram, thus explaining the Hertzsprung gap (Section 7.4). The gap, vividly seen between the giants and dwarfs at class G in Figure 3.6, broadens upward into the supergiants so as to render types F, G, and K, where evolutionary rates are particularly fast, relatively rare (Figure 12.10).

The pace of evolution then slows down, and we see stars stack up toward the red portions of their tracks, in the giants of classes K and M and the supergiants of type M. The supergiants and the ordinary giants do not evolve much later than early M: the evolutionary tracks in Figures 12.8 and 12.9 do not go much past M0, and the distribution of supergiants in Figure 12.10 stops abruptly at M4. The latest type stars – the giants that extend to M8 or even M10 – are the result of more advanced stages of evolution that we will explore below.

Clusters provide an even more dramatic illustration of how theory explains the observational data. If such a grouping is born with a full range of stellar masses, we will initially see only a main sequence. Stars will successively peel off at the top, as the earliest types will be first to evolve. As the cluster ages, we first see development of the brightest supergiants, then the dimmer ones, and finally, the ordinary giants. We first encountered this concept in Section 5.7 where we saw how clusters of differing ages could be compared to derive distance. The observed HR diagrams (Figure 12.11) can be beautifully reproduced theoretically simply by freezing the entire assembly of stars within their various evolutionary tracks at a given instant. The age of a group is readily found from the point at which the giants join the main sequence. From these studies, we determine that the globulars are the oldest of all galactic stellar systems.

Figure 12.11 provides another example of the role played in stellar evolution by chemical composition. The turnoff point of the oldest known galactic cluster, M67, is later than that of the even older globular cluster M3, which would appear to contradict the theory. We are seeing the same phenomenon that creates the subdwarfs and the horizontal branch (Sections 5.8 and 8.8). The globular cluster stars are deficient in metals, and the resulting lowered opacity of their atmospheres makes them hotter and bluer than Population I stars of the same masses would appear. Exact calculations for specific globulars (Figure 12.12), matched to the measured cluster composition, then reveal that the clusters are between 12 and 15 billion years old. Since no other older assemblies are known in the Galaxy, we assume our Milky Way system to be the same age. Moreover these deductions provide us with a powerful constraint on the age and consequent history of the Universe.

If we examine the evolution of the Galaxy as a whole through the field (non-cluster) stars, we see that the location of the common giant branch of our standard or general HR diagram (Figure 3.6) is set by the evolution of the oldest stars. There are no dimmer giants, because the Galaxy, at an age of some 13 billion years, is not old enough to have allowed stars with masses less than about 0.8 solar ever to have evolved away from the main sequence. The subgiants are seen to be only stars in transition from dwarf to giant.

As the stars evolve to the right on the HR diagram they pass through various zones in which they may temporarily oscillate. Between very roughly 2 and 10 solar masses they transit the instability strip, and appear awhile as Cepheids (Section 7.5). The lowest in this mass range become the δ Scuti stars (Sections 7.8 and 8.7). Above about

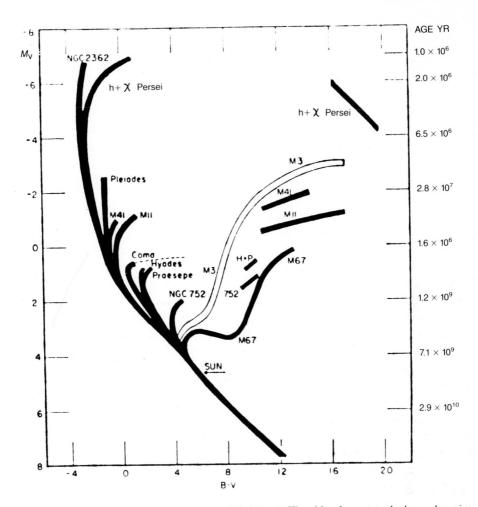

Figure 12.11. Evolution of galactic star clusters (filled lines). The older the system the lower the point between the intersection of giant branch and the main sequence. The ages for the turnoff points are given on the right. The turnoff of the oldest system, the globular cluster M3 (open line), is to the left of that of the younger galactic cluster M67 because the reduced metal content makes the Population II stars hotter and brighter. Such cluster diagrams can be theoretically reproduced by applying evolutionary tracks similar to those in Figures 12.9 and 12.10 to the main sequence. From an article by A. Sandage in the *Astrophysical Journal*.

$10 \ M_\odot$ they simply pass over the instability strip, but are so distended that they may fluctuate erratically, possess strong winds, and appear as odd or irregular variables. As the higher masses leave the main sequence, they may appear briefly as β CMa variables (Section 9.6).

The internal contraction of a giant will take millions of years, which seems like a terribly long interval, but which is rapid when compared with main sequence lifetimes. It is halted, or greatly slowed, when the temperature climbs sufficiently high to ignite the core helium. That element is very difficult to fuse and requires a nearly simultaneous

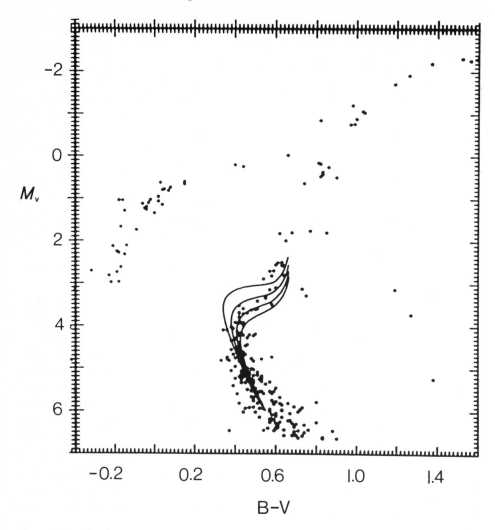

Figure 12.12. The observed and calculated distributions of stars of the famous northern globular cluster M13 in Hercules. The color–magnitude diagram is similar to that shown for M5 in Figure 8.11, except that the distance has been included so that absolute magnitude can be used instead of apparent; see also the conversion of spectral class to color in Figure 3.10. The solid lines, called *isochrones*, show the development of the lower part of the giant branch (from top to bottom) 9, 12, 15, and 18 billion years after birth. We must also adopt values of helium and metal abundance (note again the sensitivity of evolution to helium composition seen in Figure 12.8). We can clearly exclude the outer two. Fits of theory to a wide variety of globulars give an average age of about 13 billion years, which is also applied to the Galaxy as a whole. The exact value is still controversial. From an article by D. A. Vandenberg in the *Astrophysical Journal*.

collision of three atoms to produce carbon. This *triple-α process* (so called since helium nuclei are also known as *α particles*) cannot take place until the temperature reaches about 2×10^8 K. The ignition point is called the *helium flash*, and for the lower mass stars occurs at the top of the giant track, marked by the third heavy dot in Figure 12.9. Now the star settles in for another pause while the core helium is burned.

12.6 Helium burning

The life stories of the stars from this point onward depend strongly on their masses, and the helium burning pause assumes a great variety of aspects. At lower masses, under about 3 $M_\odot$ (sharp divisions are artificial, since the changes are continuous with increasing mass), the stars stabilize by moving downward, back along the red giant tracks that they had previously etched on the HR diagram. The positions subsequently taken up during fusion of core helium into carbon (because of additional reaction chains, actually into a mixture of carbon and oxygen) depend not only upon their masses, but again upon their chemical compositions. Those with solar abundances will come to rest about halfway down; the ones with lower metal abundances will, for a given mass, spread out from these locations toward higher temperatures. In a globular cluster, all the currently evolving stars are just under a solar mass; their low metallicity, plus continuous changes in luminosity and temperature as the helium is consumed, create the distinctive horizontal branch (Section 8.8), which is marked by the labeled line in the evolutionary plot. The numerous solar-type Population I stars settle into a 'clump' within the giant branch near where it would be intersected by the *HB*.

Stars with higher masses, say above 3 $M_\odot$, are really not very firmly fixed in place on the HR diagram. During helium burning they keep moving back and forth, executing 'blue loops' – brief excursions to higher temperature and back – that may take them repeatedly through the instability strip. A given star could become a Cepheid several times during its career. They also have more complex helium-burning patterns. Helium ignites in the core and burns at the helium flash (the third heavy dot in Figure 12.9), then quiets down only to enter its major phase near the end of the blueward excursion (the fourth heavy dot on the 5 $M_\odot$ and 9 $M_\odot$ curves in the figure). We see that as we proceed upward through 9 to 15 $M_\odot$ that this second zone shifts blueward, so that we can have He burning almost anywhere among the supergiants, from classes M through B. Those of very highest mass, over say 30 $M_\odot$, can actually evolve from their M supergiant phase, and then right back to O again, where, after losing considerable mass and altering their surface abundances, they might appear as Wolf–Rayet stars (Section 10.2).

The details of high-mass evolution are enormously complex, and depend critically upon many uncertain processes. All we can hope to do here is to provide something of the flavor of the subject. We will look at these brilliant stars again in Section 12.10 when we explore the supernovae.

12.7 The death of the core

We now concentrate on the fates of the lower mass stars, those below roughly 5 to 9 $M_\odot$ whose evolution creates the common giants. At some point – on the horizontal branch for Population II and within the clump for Population I – the core helium will be exhausted. The hiatus will be ended, and the central zones must once again rapidly contract with rising temperature. For the second time, since these lower masses are small enough to allow rapid communication between the core and surface, the envelope balloons outward. The star again ascends the giant branch, now with a dead degenerate carbon–oxygen core that was created from the fusion of helium. All during core helium burning, hydrogen was still being fused within a surrounding shell, adding considerably to the luminosity. With the completion of the C–O core, helium burning now spreads outward in a second shell interior to the first.

This phase of evolution is called the *asymptotic giant branch*, or *AGB*, since the second giant track approaches, or is parallel to, the first. The stars are also sometimes termed *double shell sources* because of the two zones of nuclear burning. The upward paths takes them farther to the right in the HR diagram where they are seen as the coolest giants, their evolutionary loci extending to M8 or even later. Their internal structures render them unstable, and they produce the famous Mira, or long-period, variables (Section 4.5).

The AGB path for a typical 0.55 $M_\odot$ core of solar composition is shown ascending from its helium burning position (near the right end of the HB) in Figure 12.9. We have serious problems in handling the AGB theoretically, and in predicting the precise track the star will follow. The size of the star at a given luminosity, and consequently its surface temperature, depend upon the effectiveness of internal convection in bringing fresh fuel into the nuclear burning zone. Convection is one of the most difficult of all physical problems to be dealt with by the astrophysicist. There is no viable theory for predicting the size of the convecting cells. Consequently this dimension is a free parameter that can be varied at will until we achieve the desired result, one that matches the observations: not a very satisfactory situation. The AGB track in Figure 12.9 is arbitrarily placed so as to enter the realm of the coolest stars, the Miras.

The behavior of the concentric shells of nuclear fusion around the spent core is decidedly odd and has only recently been appreciated. They do not burn together but switch on and off in alternation. At the start of the ascent of the AGB, helium first burns in a thick shell as hydrogen-burning dies away. Eventually, the helium-rich zone exhausts itself and the old hydrogen-burning shell compresses, and reignites. The fusion rate is especially fast at its lower edge where the temperature is highest, so the H-burning zone continually adds to the thickness of the quiet He-region. At some critical point the helium shell becomes thick and hot enough to fire itself and it begins to burn rapidly in what is called a *thermal pulse*. As it fuses, in the process adding the by-product of carbon to the core, it expands, pushing outward on the H-burning shell. The enlargement of this outer zone cools the gas and shuts down the hydrogen fusion. After a time the helium in the inner zone becomes depleted, which stops further activity, the hydrogen shell compresses, its temperature rises, H-burning again commences, and the whole process begins over.

The number of thermal pulses suffered by such a star depends on its mass, and the interval between them depends on both its mass and position on its evolutionary track. Typically, we might see ten or so spaced a few hundred thousand years apart. At first, the envelope of the star is so thick that the effect of the pulses, or helium flashes, is insensible at the surface. But while this complex process takes place in the shells, the AGB star begins to lose mass at a rate that climbs with increasing luminosity: recall the discussion of mass loss among the bright M giants in Section 4.6. At some point it thins enough so that we suspect that the effect of the pulse might actually be seen, although we have never firmly identified any real star with such an event. And although there are some wonderful effects yet to come, life is now nearly over.

12.8 Planetary nebulae

If the core of an evolving star, which is now defined as any part of the star that has passed through nuclear burning, remains below 1.4 solar masses, evolution is allowed to proceed quietly. This central zone will become degenerate (Section 8.9), and the pressure of its electrons will bring a final halt to the contraction that began long ago out of the interstellar medium.

The ultimate mass of the core depends upon the initial main sequence mass, which controls the rate of burning, and on the amount of the hydrogen envelope lost through winds during both red giant evolution periods. From our best estimates, an initial bulk of very roughly 8 $M_\odot$ – the kinds of stars we have been looking at – should produce a core below the critical limit. By the time the star nears the top of the AGB, it has removed all but a small fraction of its hydrogen envelope, which means that for the above 8 $M_\odot$ star, over three-fourths is ejected back into space: again, witness the immense mass loss rates we encountered for the Miras in Section 4.6.

At the end of a typical star's AGB lifetime it produces a planetary nebula (Section 11.1); the remaining core with whatever envelope it has left will be the planetary nucleus or central star (Section 11.2), and ultimately the white dwarf. The nature of this event is controversial, and we will look at two possibilities. As the star approaches the top of the AGB the thermal pulses become increasingly violent. It seems likely that the last of them (the last by definition, since it terminates this alternating kind of nuclear burning) very rapidly ejects most of the remaining hydrogen into space as an expanding shell: a nascent planetary nebula. The remnant star is a near-naked core, typically 0.6 to 0.8 $M_\odot$ (and always under 1.4 $M_\odot$), that moves very rapidly to the left on the HR diagram (illustrated by the horizontal track to the left of the AGB in Figure 12.9) as it heats, still burning some remaining hydrogen or even helium fuel in a shell around the C-O core. When the surface temperature reaches about 28 000 K, it produces enough ultraviolet radiation to illuminate the expanding planetary, which by that time is perhaps a hundredth of a parsec in radius. At first the nebula is dense and thick and absorbs all of the ionizing radiation. But as it grows and thins, it begins to leak some of this energetic starlight, and we will then see the planetary faintly surrounded by the much larger sphere of matter ejected during the earlier periods of mass loss (Figure 12.13).

Another possibility, only recently developed, is that the wind blows fairly steadily from the AGB star *without* any great increase in mass-loss rate right up to the near exposure of the core. At that point the star develops not a high *mass* wind but a high *speed* wind that compresses the circumstellar cloud into an expanding shell that will later be seen as the planetary. Evidence for this picture comes from the fast winds observed in planetary nuclei (see Section 11.3 and Figures 11.6 and 11.7) and from density enhancements in planetaries that looks like they might have been formed by

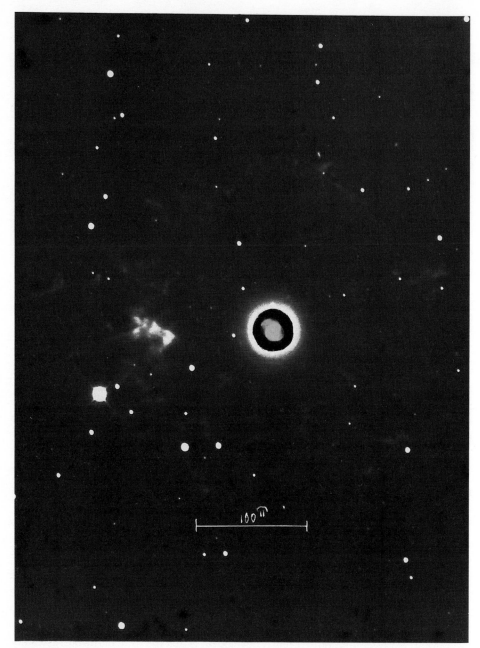

Figure 12.13. The planetary nebula NGC 6543 in Draco, set inside a huge halo that presumably is the result of the AGB mass loss that took place prior to the final ejection or creation event. The planetary is young and small and on the horizontal portion of its evolutionary track in Figure 12.14: note the brightness of the nucleus. The inner photo is a much shorter exposure than the outer. National Optical Astronomy Observatories (Kitt Peak) photos, courtesy of A. G. Millikan.

shock waves. The truth may lie in between, with a compressing shock impacting a late-developing wind, the relative importance of each phenomenon depending on the kind of planetary involved.

Whatever the scenario for planetary production we know the fate of the stellar remnant. As it comes off the AGB it moves leftward on the HR diagram at constant luminosity (typically a few thousand solar, depending on mass) to extraordinarily high effective temperatures. By the time it reaches 40 000 K or so, when the expanding nebula is still very small, it exits the standard HR diagram, or the log L–Log T plane depicted in Figure 12.9, and enters an extended version of the plane shown in Figure 12.14. Most of the star's energy is still coming from nuclear fusion that takes place between the C–O core and a small envelope. The burning eats away at the envelope from below, and fast winds dissipate it from above, causing it to thin. The hot interior

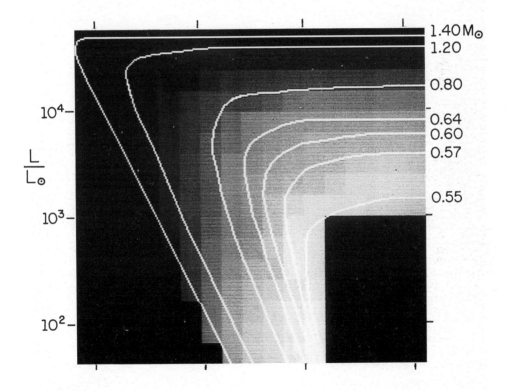

Figure 12.14. Evolutionary tracks on the log L–log T plane for planetary nuclei of different core mass, which determines the ultimate temperature reached. The track near 0.55 $M_\odot$, produced by a star of near solar mass, is an extension of that given in Figure 12.9. It exits at the bottom of this figure and reappears in Figure 12.9 at the lower left. The higher mass paths are the products of greater initial mass: the one at the top, at the white dwarf limit, derives from mid-B main sequence stars near 8 $M_\odot$. The expected numerical distribution of nuclei is shown by varying shades with the brightest indicating the most stars. They strongly cluster around the lower-mass tracks, since these come from the common and populous solar types. Cores with temperatures above 125 000 K are rare since they come from the less common upper main sequence stars, and because their rates of evolution are particularly rapid. Figure courtesy of R. A. Shaw.

Figure 12.15. YM 29, also called Abell 21, one of the larger planetary nebulae known, with a diameter of nearly a parsec. The central star, marked with the arrow, is descending along one of the tracks on the lower part of Figure 12.14, on its way becoming a white dwarf, and after several thousand years, the nebula will have expanded to invisibility. This object is highly enriched in nitrogen and helium as a result of thermonuclear processing and is the result of the evolution of a star of several solar masses. Contrast with the photo of NGC 6543 in Figure 12.13 and note that here we see a dim star within a large, old nebula. National Optical Astronomy Observatories (Kitt Peak) photo, courtesy of G. H. Jacoby.

becomes more and more exposed, and as a consequence the surface temperature climbs. Since the luminosity is constant, the star shrinks, approaching white dwarf dimensions.

The luminosity and the ultimate temperature depend on stellar mass (Figure 12.14). At typical core values of 0.6 $M_\odot$, it climbs to nearly 150 000 K; near the white dwarf limit a million degrees can be reached, but a figure this high is very rare since there are few cores this massive and their evolution is very rapid (250 000 K is near the maximum ever observed.) When the nuclear fuel begins finally to run out, the fusion engine effectively shuts down and the surface heating stops. The star now starts to cool, and since it is also still contracting, it suddenly plummets in luminosity, and now moves down and to the right on the HR diagram.

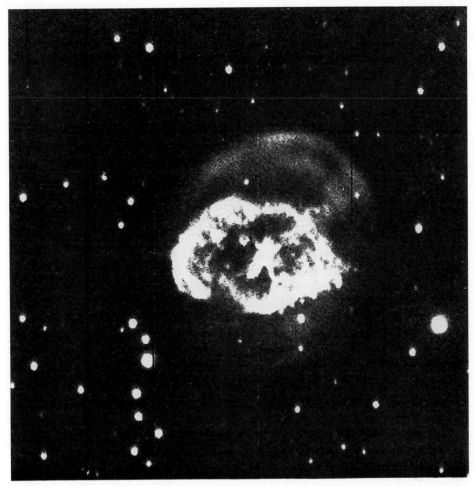

Figure 12.16. Abell 78, a large planetary nebula whose nucleus was apparently driven back to the AGB by a late thermal pulse as it was descending one of the evolutionary tracks in Figure 12.14. In this CCD image (Section 1.11) the nucleus is obscured by a small, elongated dense cloud of nearly pure helium, ejected while the star was re-visiting the giants. National Optical Astronomy Observatories (Kitt Peak) photo, courtesy of G. H. Jacoby.

All this time the planetary has been expanding; its evolution proceeds in parallel with that of its core. Consequently, we will see small nebulae a few several hundredths of a parsec in radius conjoined with bright stars (Figures 12.13 and 11.3), and large ones, a few tenths of a parsec across, coupled with faint ones (Figures 12.15, and 11.2). The core mass depends critically upon the initial mass, and of course, how much matter is lost via winds. The lowest horizontal track comes from stars born at about 0.8 $M_\odot$ (Section 12.5), and the highest, at the white dwarf limit, near our canonical (and quite uncertain) 8 $M_\odot$. It is in the larger stars that internal processes are sufficiently vigorous to cycle by-products of fusion upwards into the ejected envelope (Section 11.1). Therefore nebular abundances correlate with remnant position on the log L–log T plane. From the theory of stellar structure we can actually infer the initial mass from the resulting abundances for comparison to core mass, in order to see how much matter has actually been lost. The result of this test is in reasonable, at least qualitative, accord with the observed determinations of mass-loss rates.

The final degenerate's characteristics and spectral class – DA, DB, or other (see Section 9.10) – may at least in part be defined by the evolution that takes place during the planetary phase and maybe even on the AGB itself. For example, if we remove all of the hydrogen envelope, we may ultimately get a DB star. We do not know when the settling of helium takes place to produce the members of the DA class. But we are beginning to see this kind of separation take place as we discover that planetary nuclei also show similar strange abundances.

Before their demise, some central stars may have a last surprise waiting for us. As a planetary nucleus descends toward the white dwarfs it is still burning a bit of hydrogen, and lurking below is a quiescent helium shell. If conditions are right it can set itself off in a final thermal pulse, a curtain call that actually drives the star back to the AGB whence it re-evolves as a planetary nucleus. Buried within the much larger original nebula we can then see a tiny shell caused by the additional mass loss, this one highly enriched in the helium that earlier had lain at, or just below, the stellar surface. Only two of these bizarre characters are known (one of which is shown in Figure 12.16).

It is also possible that some stars can evolve more directly from the giant state to the white dwarfs without producing a planetary. The calculations of the birthrates of planetaries and degenerates are similar, but there is enough uncertainty to make their equality quite insecure. Moreover, we do find a number of hot stars (called subdwarf O or sdO) that may be becoming white dwarfs but that have no apparent nebulae. The actual percentage of giants that produce planetaries, although it must be large, is still unknown.

12.9 White dwarfs reprised

But even these repeaters will eventually dim out as white dwarfs, and descend along the white dwarf evolutionary loci that are outlined by the stars placed on the HR diagram in Figure 12.7. The theoretical track for a core mass of 0.6 M$_\odot$ is also shown cutting across the lower left-hand corner of Figure 12.9 after it exits Figure 12.14. Curiously, for a given temperature, the lower mass white dwarfs are the brighter: the greater gravity of those of higher mass makes them shrink more tightly, their surface areas become smaller, and so do their luminosities.

The cooling rates are now slowing dramatically. These stars live off their internal heat, and the cooler they are the more frugal they become with what energy remains. As a consequence, although the degenerates are certainly dying, none are yet what we might really call *dead*. Even the oldest still retain some heat and hang on to the HR diagram below the main sequence K and M stars. We have not been able to predict very well just where the end point should be, given the age of the Galaxy. It will, however, be billions of years into the future before one at last disappears from our view as a cold cinder.

Figure 12.17. A supernova in the galaxy M63 (NGC 5055) in Canes Venatici. In spite of the galaxy's distance of four and one-half million parsecs, this Type I supernova appeared here on June 22, 1971 at magnitude 13.5. In early June it was 11th magnitude and visible in a small telescope, when it had an absolute magnitude of −17. University of Illinois Prairie Observatory photograph.

In the broadest sense, the end product is determined by the initial characteristics of youth on the main sequence. In principle we should be able to follow a star with our theories from the time it makes its first appearance to when it dies as a white dwarf. But we cannot yet make such detailed predictions. In spite of our great apparent knowledge, we can trust our theories over only short segments of the evolutionary tracks. There are still too many uncertainties to allow us to tie them all together.

12.10 Violent eruptions: supernovae

The more massive stars, those that do not produce planetaries and white dwarfs, have a very different fate. If they cannot lose enough matter to get below the Chandrasekhar limit (the highest allowable white dwarf mass, Section 8.9), they may explode in a grand outburst that can be seen millions of parsecs away, with absolute magnitudes between −17 and −20. We unfortunately know very little about the causes of these *supernovae*; many of the processes are still beyond our understanding. The problem is compounded by our realization that some lower mass stars might also produce supernovae if their usually lighter-weight cores can somehow be teased into exceeding the limit. The remnants of such explosions, if they exist at all, are far too strange to place on the HR diagram however it is constructed or altered from the original. But for the sake of completeness, we must look at some of the possibilities for stellar catastrophe and what these epic blasts might yield.

Compared to other kinds of celestial objects, observations are scanty. We have not witnessed a supernova in our Galaxy since Kepler's and Tycho's stars of 1604 and 1572, both of which predate the telescope. Most of what we know comes from galactic remnants of the detonations and of direct observation of these events in other, usually distant, galaxies (Figure 12.17), which renders observation very difficult in spite of the immensity of the explosions. It was for this reason that the astronomical world was so electrified by the supernova that took place in the Large Magellanic Cloud in 1987 (Figure 12.18), which allowed us finally to study one close-up with modern observing techniques. This spectacular event was easily visible to the naked eye as a third magnitude star near the Tarantula Nebula (Figure 10.15), even though 50 000 parsecs distant. Of greatest importance was our ability to find for the first time the nature of the star that exploded, which provides a fundamental constraint on theory.

A standard picture of a supernova outburst involves straightforward evolution of single massive stars. Somewhere between 10 and 20 solar masses and upward, the structures of the stars are such that their carbon cores do not become degenerate. They can therefore exceed the white dwarf limit without collapsing. In addition, the envelopes are so large to start with that they retain considerable mass. We think that some of these can be seen at certain points in their development as the Wolf–Rayet stars (Sections 10.2 and 12.6), as mass loss peels away their outer skins to reveal processed matter that has been cycled upwards.

As evolution proceeds, these great stars pass through a succession of nuclear burning stages and evolutionary pauses. A massive star's large envelope allows the internal temperature to keep climbing. The progression of hydrogen to helium and then helium to carbon (or carbon and oxygen) can then continue. The C–O core that developed from helium burning contracts, and when the heat is sufficient, begins to fuse to magnesium. When that fire expires, the core contracts again, now surrounded by *three* shells, in outward succession burning carbon, helium, then hydrogen. The sequence continues through neon fusion, then through that of silicon, up to the formation of an iron core surrounded by burning shells of silicon, neon, carbon, helium, and hydrogen. But iron is the end; it cannot be an energy source, and once contraction sets in, further nuclear fusion cannot stop it. The center contracts with great speed, and the temperature continues upwards past the ten billion degree mark. So much energy is created by the collapse that the iron nuclei break down. A blast wave proceeds through the outer layers causing sudden additional nuclear burning there, and a supernova erupts. Exactly where on the HR diagram this event should occur, and how it relates to our theoretical evolutionary tracks is yet something of a mystery: possible progenitors include red supergiants, Wolf–Rayet stars, objects like η Carinae (Section

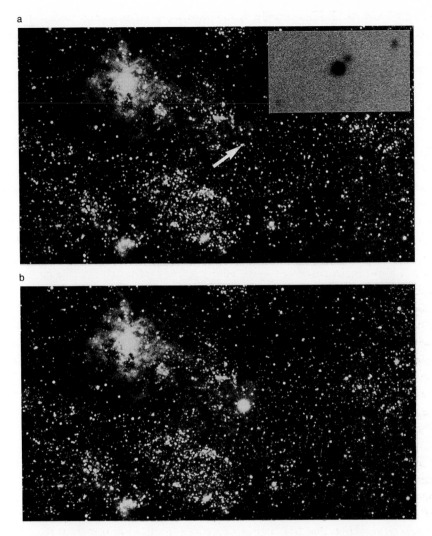

Figure 12.18. Supernova 1987A in the Large Magellanic Cloud. (*a*) shows a pre-outburst view of the section in the Large Magellanic Cloud in which the supernova appeared. The Tarantula Nebula (Figure 10.15) is at the upper left. The arrow points to Sanduleak −69° 202, the B3 Ia supergiant that exploded into SN 1987A seen in (*b*). The inset at the upper right shows a vastly magnified view of the pre-outburst star taken with the Cerro Tololo 4 meter telescope. The bulge at the lower left of the bright supergiant was for a time considered a candidate, but we now know that the event was caused by Sk −69° 202 itself. The star is now *gone*. Large photographs: Mt. Wilson, and Las Campanas Observatories, Carnegie Institution, courtesy of the University of Toronto and Ian Shelton; inset: National Optical Astronomy Observatories (CTIO), courtesy of You-Hua Chu.

11.7) and, with the advent of SN 1987A, ordinary B supergiants. We are far from mastering this demanding science.

12.11 Neutron stars and pulsars

The collapsing core now becomes degenerate, and since it is more massive than a white dwarf hurtles past the radius at which contraction can be stopped by electron degeneracy. The densities become so high that electrons and protons are merged together into neutrons. At a tiny diameter of only ten or so kilometers, one thousand

Figure 12.19. Cygnus X-1, a powerful X-ray source, is identified with a massive B star, which appears to revolve around an invisible companion. The orbital characteristics indicate that the missing star is massive enough to be that ultimate product of stellar evolution, the *black hole*, from which no radiation can escape. The X-rays are caused by gas flowing from the B star being heated in an accretion disk around the hole. © National Geographic–Palomar Sky Survey, reproduced by permission of the California Institute of Technology, courtesy of C. T. Bolton.

times smaller than a white dwarf, packing some 10^{12} grams per cubic centimeter, the neutrons become degenerate, and bring the collapse to a halt.

After the explosion is over and the debris has cleared away, some of these *neutron stars* are eventually observed as *pulsars*: oblique rotators (stars whose rotational and magnetic axes are not aligned, see Section 8.4) that spray radiation outward from the poles of their powerful magnetic fields. These have been squeezed with the matter to strengths of some 10^{12} gauss (compare with the Sun's one to two gauss and the 'mere' megagauss fields of some white dwarfs that we discussed in Section 9.10.) If a young neutron star is properly positioned so that during the spin one of its magnetic poles sweeps across the direction to Earth, we will see (in all but a few cases only in the radio spectrum) a pulse of radiation once every rotational period, which is typically one to a few seconds. An optically visible pulsar is illustrated in the inset of Figure 12.21 below.

Like their distant white dwarf cousins, neutron stars are also limited by mass. If they exceed about four times solar, even degenerate neutrons are insufficient to hold back gravity, and the star – if we can still call it that – collapses further into a black hole, which is so dense that no radiation can escape its gravitational grasp. We do not yet understand what kind of star is destined to be one of these strange bodies. Perhaps they are the result of supernovae; possibly a very massive star could contract into one without an explosion. We have good evidence, however, that they are there, as Cyg X-1 (Figure 12.19), and LMC X-3 in the Large Magellanic Cloud, attest. From these bodies we observe X-rays that seem to be generated as matter flows from a binary companion and heats as it accretes into a disk (Sections 11.6, 11.9, Figure 11.10) and then falls into the black hole.

The in-between masses – 8 to 12 or so times solar – present a problem. They are too large to develop into white dwarfs: the growing degenerate C–O core, which gathers matter from the He-burning shell, becomes more massive than the Chandrasekhar white dwarf limit and collapses. There is no clear indication as to whether these produce supernovae, or non-explosively develop into neutron stars.

12.12 Varieties of supernovae

Our observations of the light curves and spectra of supernovae show us that there are two distinct kinds, called simply *Type I* and *Type II*. The former exhibit no hydrogen lines during the explosion, whereas the latter clearly do: the chemical compositions of the progenitors seem to be quite different. The light curves are also dissimilar: the Type II variety is the fainter, near the bottom of the range, and often shows a broad plateau after maximum that is absent in the curves of the others, which decline steadily. The dimming rate of the typical type I outburst in fact closely matches what we would expect from the decay of the isotope nickel 56 into cobalt 56, thence to iron 56, showing nucleosynthesis in action.

The Type II are found in the spiral arms of galaxies where they are rather clearly associated with the fates of the massive stars described above, and therefore are confusingly associated with Population I. The Type I are generally found in elliptical galaxies of Population II (never look for consistency of nomenclature in a growing field of study), in galactic disks, and in irregular galaxies that undergo active star formation. It appears certain that at least some of the Type I explosions cannot come from massive stars, since they usually occur in elliptical galaxies in which the upper main sequence long since burned away. We think that binaries may be the culprits. A white dwarf in close enough orbit about an evolving star (Section 11.4) might accrete enough mass to grow larger than the Chandrasekhar limit and explode. Ordinary novae are produced this way, though, and repeated smaller detonations should keep the mass too low for such an event to occur. A newer idea is that Type I eruptions arise from double white dwarfs. If they are sufficiently near one another, they will spiral together, make contact, coalesce, and explode. There is very little hydrogen left to start with so none will be seen in the resulting spectra of the event. Some of these detonations may wipe out the stars altogether, leaving no stellar remnants at all.

The more we observe, the more complicated the situation becomes. For example, we now recognize a subdivision of Type I, called Ib, which appears in spiral arms, but which like ordinary Type I lacks hydrogen lines. Perhaps the supernovae of this set represent explosive finales of the Wolf–Rayet stars, which previously lost their hydrogen envelopes via winds, or members of binaries that have had their outer envelopes stripped away by mass transfer. We certainly have not seen the end of the discovery of different kinds of events as our data improve. As an example, keep reading.

12.13 SN 1987A

Early on the morning of February 24, 1987 an astronomer named Ian Shelton of the University of Toronto took a routine photograph of the Large Magellanic Cloud (the *LMC*) with the Carnegie Institution's 10-inch Bruce telescope at Las Campanas, Chile. Upon developing his plate Shelton noticed a fifth magnitude star that had not been there a day earlier. Since the last supernova that even approached being nearby, the one that occurred in the heart of the great Andromeda galaxy M31 (Figure 1.9) in 1885 (and which almost reached naked eye visibility), astronomers have been avidly anticipating another close one that could be studied with an increasing arsenal of modern instrumentation. And now one had arrived (Figure 12.18).

But if we had expected simple confirmation of the existing theories, we were wrong: SN 1987A confounded us with its odd behavior, and thereby was teaching us something new. At the position of the brilliant event was a 12th magnitude B3 supergiant known only by its catalogue names of Sanduleak −69° 202 and Cape Photograph Durchmusterung, CPD, −69° 402. But if that had been the progenitor, its climb had been a 'mere' seven magnitudes at discovery, a total only of nine by maximum, culminating in an absolute magnitude of −15.5 by late April. So the object was rather faint, since Type II supernovae, which this one proved to be upon development of its powerful hydrogen lines (Figure 12.20), should reach at least to −17 or so. Moreover, the star took a long time to get there, brightening through all of March, April, and even, to a small degree, into May. Thus, the light curve was decidedly odd, as it kept climbing when it was expected to be sharply dimming.

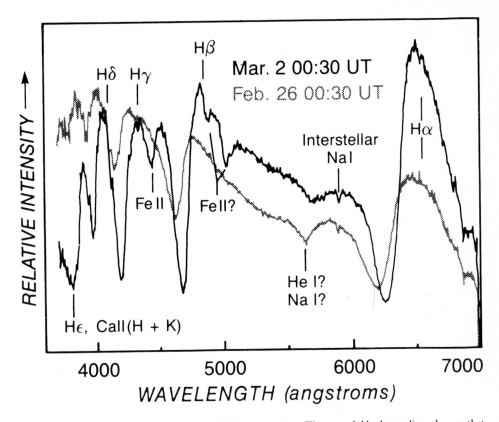

Figure 12.20. Optical spectra of Supernova 1987A on two dates. The powerful hydrogen lines show us that it is a Type II event, which is related to the stars of Population I. The Hα and Hβ lines show P Cygni profiles (Section 9.5) indicative of outflowing gas. Mt. Wilson and Las Campanas Observatories, Carnegie Institution of Washington, spectra courtesy of *Sky and Telescope*, M. Gregg, R. A. Kimble and A. Davidson.

Odder still was the spectral type of the progenitor: we might expect the highly evolved Wolf–Rayet stars, or M-type supergiants to explode, but not a normal-seeming one of class B. So for a time we thought that the Sk −69° 202 might be just a foreground object, with the *real* former star hidden behind it. The faint bulge on the B star seen at the lower left of its image (see the inset in Figure 12.18) was an especially good candidate. But as positional information improved, we were able to exclude the others: it was the B star after all.

If we compare the position of Sk −69° 202 in Figure 12.7 with the evolutionary tracks in Figure 12.10, after application of the bolometric correction, we would estimate a mass in the neighborhood of 20 Suns (not entirely apt, since these tracks are for stars of normal composition, not the decreased metallic abundance of the LMC), which is in the lower part of the allowed range for Type II core collapse events, but still appropriate. That such a collapse indeed took place comes from a remarkable and unique set of observations. Two massive neutrino detectors, one in Japan and one in the United States, had been built somewhat earlier in order to determine the decay rate of the proton. Now neutrinos are expected to be produced in vast numbers as a result of intense thermonuclear reactions set off during the collapse, and they leave the core of the star immediately, unimpeded by the overlying layers. In fact the vast majority of the energy of such a supernova is carried away by them: the optical event, as great as it may seem, is only a ghost of what is really taking place. At 7:36 Universal Time, February 23, 21 hours before Shelton's discovery, and 5 hours before it was recorded on the rise by Robert McNaught in Australia, these detectors recorded a burst of 19 neutrinos – about what would be expected. So not only do some parts of our theories work, but incredibly, we know to within a minute when the supernova began *deep inside a star 50 000 parsecs away.*

We think that at least some of the oddness of this star is caused by the lowered metallicity of the LMC. Instead of blowing up as an M supergiant, as one might in our own Galaxy, the different composition makes the event happen while the star is blue: that is, Rigel and its ilk will not explode. The fewer metal atoms also seem to account for the star's relative faintness. Supernovae such as this one may well not be so unusual after all – we just have not recorded them since they are so dim. We will be years, even decades, in sorting out the data, and applying theories, to them; as we wait in even greater anticipation of the next one.

12.14 Supernova remnants

Whatever their origins, supernova outbursts are responsible for most of the heavy elements in the Universe. The eruptions spew forth the products of the previous burning stages, and the very high temperatures cause additional nuclear reactions that can create atoms as heavy as uranium. We observe the result of the detonations in the expanding clouds of exploded gases called *supernova remnants* (the term is never used for the stellar by-product), of which the Crab Nebula (Figure 12.21) is the most readily visible example. In the Crab we see highly enriched matter, mostly helium, which will eventually blend with and enrich the gases of interstellar space. In some instances, as in the case of the famed Cygnus Loop (Figure 12.22), we see only the blast wave itself as it sweeps up and heats the interstellar medium.

A large number of these remnants are known, many heavily obscured by interstellar dust and visible only with radio telescopes. Although they may look vaguely like H II regions (Section 10.5), much of their radiation is produced very differently, by free electrons moving near the speed of light accelerating in a magnetic field: a process called the *synchrotron* mechanism. Without question these graceful forms are among the loveliest the sky has to offer: beauty from catastrophe.

The supernovae present us with many puzzles. We do not, for example, see a good match between pulsars and remnants: most pulsars have no associated gas clouds, and most remnants have no pulsars. Perhaps these odd stars outlive their expanding ejecta, and pulsars inside many remnants are simply oriented in the wrong direction. As pointed out above, the relations between initial stellar mass, population type, binary

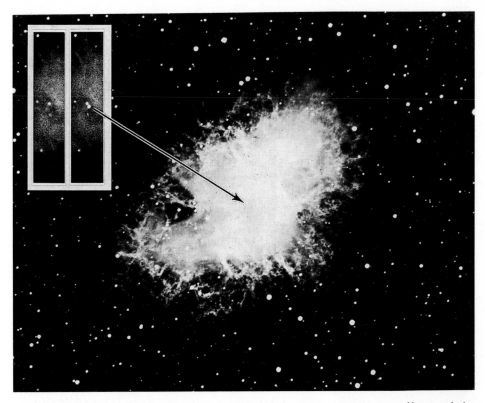

Figure 12.21. The Crab Nebula, the most famous of all gaseous supernova remnants, caused by an explosion easily visible on Earth in the year 1054. The stellar core, left behind after the explosion, is now a dense neutron star and powerful pulsar, one of the few that is associated with a gaseous remnant. The Crab pulsar, only about 10 kilometers across, is one of the fastest known, turning on and off 30 times per second, implying a rotation period of only 0.03 seconds. It can be seen in the inset at the upper left: the left side of the inset shows the interior of the nebula between pulses, and no trace of the neutron star; the right side shows it at maximum emission. As it ages it will stop radiating optically and will be detected only as a radio source, more typical of pulsars. National Optical Astronomy Observatories (Kitt Peak) photographs, the large photo courtesy of A. G. Millikan.

incidence, explosion mechanism, and end product is very confused. And the recent nearby event, SN 1987A, which we had hoped would clarify the picture, has just confounded us even more. In addition, the theories predict supernovae at such a rate that the remnants should have fed much more iron into the Galaxy than we actually see in young stars. In spite of our seeming sophistication, we still do not know just how it is that stars explode.

Nevertheless, even if we do not yet understand all the numbers and the details, it is clear that these grand explosions provide the principal means by which the Milky Way is enriched over time with metals. Largely because of them, the ancient metal-poor main sequence defined by the subdwarfs has shifted to the right on the HR diagram to

a   b

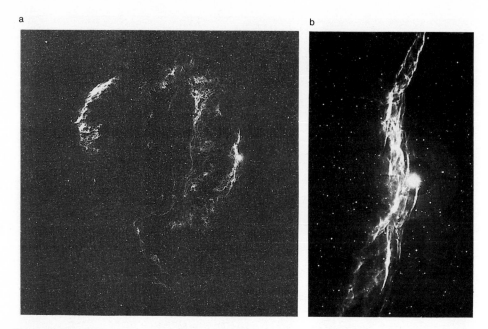

Figure 12.22. (*a*) The Cygnus Loop (the Veil Nebula), a large old supernova remnant nearly 3° across, whose star has never been identified. We see here not the gas ejected by the explosion, but interstellar matter heated by a shock wave. A high resolution photograph of the right-hand arc taken in the light of the [O III] lines (Section 10.7) is shown in (*b*), to illustrate the exquisite filamentary structure that is produced by folded and convoluted sheets of gas. The bright star is 52 Cygni, a K0 giant. (*a*) © 1960 National Geographic–Palomar Sky Survey, reproduced by permission of the California Institute of Technology; (*b*) University of Michigan photograph, courtesy of W. Blair.

yield the current main sequence of the solar neighborhood, which with its evolved stellar products – the giants, supergiants and all the others – has been the focus of this story.

These extraordinary finales to the stars also mark the end of our journey through the dense, complex forests and fields of stars and stellar types. So now let us turn our thoughts away from rare, sudden, violent events to the quiet, vastly more gentle view of the heavens explored throughout these pages. What a marvelous path we have followed from the first faint beginnings by Fraunhofer and Secchi, past the work of Pickering, Fleming, Maury, and Cannon, through that of Morgan, Keenan, and Kellman, and on into the burgeoning knowledge of our own times. Before leaving, look again across the HR diagram presented here, and ponder once more this remarkable array of stars. And tonight, if it is clear, go out to examine the real thing: all the classes arrayed for you, splashed wondrously across the darkened sky.

Star index

Stars are arranged by constellation, with other star names interspersed alphabetically. Within a constellation the Bayer letters are given first, followed by Flamsteed numbers, variable stars arranged in traditional order, and then other names that take on the genetive form. Stellar spectra are indicated with an asterisk.

Subject index

A stars, 165ff
  giants, 170
  metallic-line (Am), 172
  spectra, 168
  white dwarfs, 180
Abell 78, 273
absolute magnitudes, 24
absorption lines, 40, 42, 48
accretion
  in novae, 235
  in symbiotics, 245
accretion disk, 237, 240
activity cycle, 138ff
  of stars, 141
AGB stars; *see* asymptotic giant branch
    stars
age of galaxy, 266
aging of stars, 29, 101, 122, 144
  lithium, 144, 258
  magnetic braking, 155
  O and B stars, 184
AI Velorum stars, 162
alpha particles, 36
Am stars, 173, 237
Andromeda, 15
Ångstrom, 21
angular momentum, 249
Ap stars, 173
apparent magnitudes, 22
Argo, 7
Astronomical unit, 10
asymptotic giant branch stars, 261, 268
  production of planetary nebula, 269
  thermal pulse in, 268
atom, 31ff
atomic collisions, 43, 169
atomic nucleus, 31
atomic number, 32

atomic weight, 32

B stars, 183ff
  giants, 193
  spectra, 188
  white dwarfs, 198
B − V, 26
  relation to temperature and spectrum, 80
  relation to U − B, 80, 197
Balmer continuum, 42
Balmer series, 38, 39, 49
  of helium, 44
band head, 47
Bappu, V., 113, 114
Barium stars, 116
Barnard 86, 196, 249
Bayer, Johann, 5
Be stars, 191ff
Bessel, Friedrich, 118
Beta Canis Majoris stars, 193
  evolution of, 265
Beta Cephei stars, 193
Beta rays, 36
Big Dipper, 5, 9, 10, 12, 13, 165
binary stars, 11
  derivation of masses, 114
  eclipsing, 114, 115
  mass transfer, 235, 237
  spectroscopic, 50
  visual, 114
bi-polar nebulae, 252, 255
birth of stars, 249ff
blackbody, 26, 40, 41, 44
black holes, 237, 246, 278
blue loops, 267
Bok, Bart J., 196
Bok globule, 196
bolometric corrections, 92, 204